Latent Semantic Mapping: Principles & Applications

Latent Semantic Mapping: Principles & Applications
Jerome R. Bellegarda

ISBN: 978-3-031-01428-4   paperback
ISBN: 978-3-031-01428-4   paperback

ISBN: 978-3-031-02556-3   ebook
ISBN: 978-3-031-02556-3   ebook

DOI: 10.1007/978-3-031-02556-3

A Publication in the Springer series

SYNTHESIS LECTURES ON SPEECH AND AUDIO PROCESSING #3

Lecture #3

Series Editor: B.H. Juang, Georgia Tech

Library of Congress Cataloging-in-Publication Data

Series ISSN: 1932-121X   print
Series ISSN: 1932-1678   electronic

First Edition
10 9 8 7 6 5 4 3 2 1

Latent Semantic Mapping: Principles & Applications

Jerome R. Bellegarda
Apple Inc.

SYNTHESIS LECTURES ON SPEECH AND AUDIO PROCESSING #3

ABSTRACT

Latent semantic mapping (LSM) is a generalization of latent semantic analysis (LSA), a paradigm originally developed to capture hidden word patterns in a text document corpus.

In information retrieval, LSA enables retrieval on the basis of conceptual content, instead of merely matching words between queries and documents. It operates under the assumption that there is some latent semantic structure in the data, which is partially obscured by the randomness of word choice with respect to retrieval. Algebraic and/or statistical techniques are brought to bear to estimate this structure and get rid of the obscuring "noise." This results in a parsimonious continuous parameter description of words and documents, which then replaces the original parameterization in indexing and retrieval.

This approach exhibits three main characteristics:

- *discrete* entities (words and documents) are mapped onto a *continuous* vector space;

- this mapping is determined by *global correlation patterns*; and

- *dimensionality reduction* is an integral part of the process.

Such fairly generic properties are advantageous in a variety of different contexts, which motivates a broader interpretation of the underlying paradigm. The outcome (LSM) is a data-driven framework for modeling meaningful global relationships implicit in large volumes of (not necessarily textual) data.

This monograph gives a general overview of the framework, and underscores the multi-faceted benefits it can bring to a number of problems in natural language understanding and spoken language processing. It concludes with a discussion of the inherent tradeoffs associated with the approach, and some perspectives on its general applicability to data-driven information extraction.

KEYWORDS

natural language processing, long-span dependencies, data-driven modeling, parsimonious representation, singular value decomposition.

Contents

List of Figures

List of Tables

Part I

Principles

CHAPTER 1

Introduction

Over the past few years, *latent semantic mapping* (LSM) has emerged as a powerful, data-driven framework for modeling global relationships implicit in large volumes of data. Though now applied to a wide range of applications, its initial incarnation, known as *latent semantic analysis* (LSA), was specifically concerned with textual data.[1] This chapter starts with a modicum of historical perspective on the matter.

1.1 MOTIVATION

Originally formulated in the context of information retrieval, LSA arose as an attempt to improve upon the common procedure of matching words of queries with words of documents [26, 58, 64, 69, 84, 85, 102, 103, 119, 140]. The problem is that users typically retrieve documents on the basis of conceptual content, and individual words provide unreliable evidence about the conceptual topic or meaning of a document. There are usually many ways to express a given concept, so the literal terms in a user's query may not match those of a relevant document. This tends to decrease the "recall" performance of retrieval systems. In addition, most words have multiple meanings, so the literal terms in a user's query may match terms in documents that are not of interest to the user. This results in a poorer "precision" performance. Formal definitions of these retrieval evaluation measures can be found in Table 1.1.

The LSA paradigm operates under the assumption that there is some underlying latent semantic structure in the data, which is partially obscured by the randomness of word choice with respect to retrieval. Algebraic and/or statistical techniques are brought to bear to estimate this latent structure and get rid of the obscuring "noise." The outcome is a parsimonious continuous parameter description of terms and documents based on the underlying structure. This low dimensionality description then replaces the original parameterization in indexing and retrieval [26, 64].

In practice, the "latent structure" is conveyed by correlation patterns, derived from the way individual words appear in documents: this entails an elementary "bag-of-words" model

---

[1]In the specific context of information retrieval, LSA is also often referred to as *latent semantic indexing* (LSI). In this monograph we use LSA rather than LSI for consistency of terminology.

TABLE 1.1: Information Retrieval Terminology

In information retrieval, two widely used evaluation measures are *recall* and *precision*:

$$\text{recall} = \frac{\text{number of relevant documents retrieved}}{\text{total number of relevant documents}}$$

$$\text{precision} = \frac{\text{number of relevant documents retrieved}}{\text{total number of retrieved documents}}$$

Recall thus shows the ability of a retrieval system to present all relevant items, while precision shows its ability to present only relevant items. In other words, recall reveals any "leaks," and precision any "noise," in the retrieval procedure.

of the language. Moreover, in line with the data-driven nature of the approach, "semantic" merely implies the fact that terms in a document may be taken as referents to the document itself or to its topic. These simplifications notwithstanding, there is abundant evidence that the resulting description is beneficial: it indeed takes advantage of implicit higher order structure in the association of terms with documents, and thereby improves the detection of relevant documents on the basis of terms found in queries [102, 103]. It is well documented, for instance, that LSA improves separability among different topics, cf. the simple information retrieval example of Fig. 1.1.

1.2   FROM LSA TO LSM

The success of LSA in information retrieval led to the application of the same paradigm in other areas of natural language processing, including word clustering, document/topic clustering, large vocabulary speech recognition language modeling, automated call routing, and semantic inference for spoken interface control [8, 9, 10, 19, 22, 39, 48, 75, 78]. The application to clustering, for instance, follows directly from the improved separability just discussed.

To illustrate it at the word level, Table 1.2 lists two typical LSA word clusters obtained using a training collection of about 20,000 news articles from the *Wall Street Journal* database, cf. [19]. To illustrate it at the document level, Fig. 1.2 lists the distributions of four typical hand-labeled topics against 10 LSA document clusters, obtained using a training collection of about 4000 heterogeneous articles from the *British National Corpus* database, cf. [78]. In both cases, clustering was performed using standard clustering algorithms (cf., e.g., [7]) operating in the LSA space.

What these solutions have in common is that they all leverage LSA's ability to expose global relationships in the language in order to extract useful metadata regarding topic context

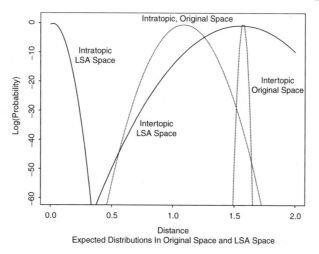

Expected Distributions In Original Space and LSA Space

FIGURE 1.1: Improved topic separability under LSA (after [119]). Consider an (artificial) information retrieval task with 20 distinct topics and a vocabulary of 2000 words. A probabilistic corpus model generates 1000 documents, each 50 to 100 words long. The probability distribution for each topic is such that 0.95 of its probability density is equally distributed among topic words, and the remaining 0.05 is equally distributed among all 2000 words. How does the average distance between documents evolve upon exploiting the LSA paradigm?

To answer this question, a suitable version (see Chapter 3) of the distance between all pairs of documents is calculated both before ("original space") and after ("LSA space") applying LSA. This leads to the expected distance distributions below, where a pair of documents is considered "Intra-Topic" if the two documents were generated from the same topic and "Inter-Topic" otherwise. Using LSA does not appreciably affect the average distance between intertopic pairs, but dramatically reduces the average distance between intratopic pairs. In addition, the standard deviation of the intratopic distance distribution also becomes substantially smaller. As a result, separability between intra- and intertopic pairs is much better under LSA, despite a sharp increase in the standard deviation of the intertopic distance distribution.

and meaning, and thereby make progress toward more intelligent human–machine communication. More specifically, three unique factors seem to make LSA particularly attractive:

- the mapping of discrete entities (in this case words and documents) onto a continuous parameter space, where efficient machine learning algorithms can be applied;

- the dimensionality reduction inherent in the process, which makes complex natural language problems tractable; and

- the intrinsically global outlook of the approach, which tends to complement the local optimization performed by more conventional techniques.

TABLE 1.2: Two Typical LSA Word Clusters (cf. [19])

Cluster 1: Andy, antique, antiques, art, artist, artist's, artists, artworks, auctioneers, Christie's, collector, drawings, gallery, Gogh, fetched, hysteria, masterpiece, museums, painter, painting, paintings, Picasso, Pollock, reproduction, Sotheby's, van, Vincent, Warhol

Cluster 2: appeal, appeals, attorney, attorney's, counts, court, court's, courts, condemned, convictions, criminal, decision, defend, defendant, dismisses, dismissed, hearing, here, indicted, indictment, indictments, judge, judicial, judiciary, jury, juries, lawsuit, leniency, overturned, plaintiffs, prosecute, prosecution, prosecutions, prosecutors, ruled, ruling, sentenced, sentencing, suing, suit, suits, witness

Note that these word clusters comprise words with different part of speech, a marked difference with conventional class *n*-gram techniques (cf. [115]). This is a direct consequence of the semantic nature of the derivation. Second, some obvious words seem to be missing from the clusters: for example, the singular noun "*drawing*" from cluster 1 and the present tense verb "*rule*" from cluster 2. This is an instance of a phenomenon called *polysemy*: "*drawing*" and "*rule*" are more likely to appear in the training text with their alternative meanings (as in "*drawing a conclusion*" and "*breaking a rule*," respectively), thus resulting in different cluster assignments. Finally, some words seem to contribute only marginally to the clusters: for example, "*hysteria*" from cluster 1 and "*here*" from cluster 2. These are the unavoidable outliers at the periphery of the clusters.

It is through these three characteristics that LSA indirectly manages to uncover some useful aspects of the semantic fabric of the document, even as they are otherwise masked by the surface forms, in this case the words.

These are, of course, fairly generic properties, which are desirable in a variety of different contexts. Categorical problems are notoriously hard to model statistically, especially when they comprise an extremely large number of categories [130]. Dimensionality reduction is routinely critical to uncovering the important structural aspects of a problem while filtering out "noise" [103]. And a global outlook can often reveal discriminative characteristics which are essentially impossible to recognize locally [8].

These observations have in turn sparked interest in several other potential uses of the basic paradigm, cf. [13–18, 21]. In such applications, the task at hand may not be as directly language-related, but the approach has nevertheless proved very appealing. Hence the change of terminology to *latent semantic mapping* (LSM), to convey increased reliance on the general

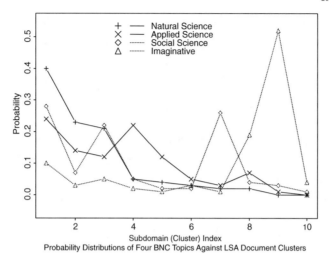

Probability Distributions of Four BNC Topics Against LSA Document Clusters

FIGURE 1.2: Typical hand–labeled versus LSA topics (after [78]). While not perfectly matching the hand-labeling, LSA-based document clustering seems reasonable. In particular, as one would expect, the distribution for the natural science topic is relatively close to the distribution for the applied science topic (cf. the two solid lines), but quite different from the two other topic distributions (in dashed lines). From that standpoint, the data-driven LSA clusters appear to adequately cover the semantic space.

properties listed above, as opposed to a narrower interpretation in terms of specific topic context and meaning.

1.3 ORGANIZATION

The purpose of this monograph is to review such properties, and some of their applications, under the broad perspective of data-driven information extraction. As the title implies, the book is organized into two main parts, one centered around the general principles underlying LSM, and the other articulating around some specific applications where these principles can be exploited. The final part of the monograph provides a discussion of LSM's general applicability and associated perspectives on the matter.

1.3.1 Part I

In addition to the present introduction, Part I comprises Chapters 2 through 5:

Chapter 2 gives an overview of the mechanics of LSM, and in particular how to automatically achieve latent semantic embedding from empirical co-occurrence data using singular value decomposition.

Chapter 3 discusses the salient characteristics of the resulting LSM feature space, including the kind of closeness measures that can be defined in that space, and how the mapping works for input data not seen in the training corpus.

Chapter 4 addresses some of the computational issues that arise in the LSM derivation, for both (offline) training and (online) mapping. This chapter also describes potential computational shortcuts and corresponding tradeoffs.

Chapter 5 considers possible probabilistic interpretations of LSM, and ensuing implications for parameter training. We also comment on the necessary mathematical assumptions and associated practical limitations.

1.3.2   Part II

Part II comprises Chapters 6 through 11:

Chapter 6 describes how to leverage the LSM framework for the specific purpose of junk e-mail filtering. We show how this approach compares favorably to state-of-the-art machine learning solutions.

Chapter 7 applies LSM results to more generic semantic classification, typified here by desktop interface control. We go over the resulting semantic inference paradigm and illustrate it on a simple example.

Chapter 8 turns to the area of statistical language modeling, where multispan language modeling can be achieved by incorporating an LSM component into a standard n-gram language model.

Chapter 9 tackles an application where LSM is exploited outside of the word-document setup. We consider the problem of pronunciation modeling, for which the mapping is applied at the character level.

Chapter 10 expands the focus to a more general pattern recognition problem. We look at LSM as a feature extraction mechanism, with application to speaker verification.

Chapter 11 further generalizes the framework to a signal processing environment, where the LSM feature space is used as an alternative transform domain. This is done in the context of text-to-speech (TTS) synthesis.

1.3.3   Part III

Finally, the last two chapters conclude the monograph:

Chapter 12 discusses the inherent tradeoffs associated with LSM, and assesses its general profile of applicability for data-driven information extraction across a variety of fields.

Chapter 13 provides a summary of the material covered in the book, as well as final perspectives on latent semantic mapping.

CHAPTER 2

Latent Semantic Mapping

Let $\mathcal{M}$ be an inventory of M individual units, such as words in some underlying vocabulary, and $\mathcal{N}$ be a collection of N "meaningful" compositions of said units, such as documents in some relevant text corpus. While the order of M and N varies greatly with the application considered, it usually lies between 1000 and 100,000; the collection $\mathcal{N}$ might comprise a total number of units in the hundreds of millions. The LSM paradigm defines a mapping between the *discrete* sets $\mathcal{M}$, $\mathcal{N}$ and a *continuous* vector space $\mathcal{L}$, whereby each unit r_i in $\mathcal{M}$ is represented by a vector $\bar{u}_i$ in $\mathcal{L}$, and each composition c_j in $\mathcal{N}$ is represented by a vector $\bar{v}_j$ in $\mathcal{L}$.

2.1 CO-OCCURRENCE MATRIX

The starting point is the construction of a matrix (W) of co-occurrences between units and compositions. In line with the semantic nature of the approach [102], this construction normally disregards collocational information in unit strings: the context for each unit essentially becomes the entire composition in which it appears. Thus, the matrix W is accumulated from the available training data by simply keeping track of which unit is found in what composition.

This accumulation involves some suitable function of the number of times each unit appears in each composition. Although many such functions are possible (cf. [63, 64]), it is often desirable to normalize for composition length and unit entropy [10], in which case the (i, j) cell of W is obtained as

$$w_{i,j} = (1 - \varepsilon_i) \frac{\kappa_{i,j}}{\lambda_j}, \qquad (2.1)$$

where $\kappa_{i,j}$ is the number of times r_i occurs in c_j, λ_j is the total number of units present in c_j, and ε_i is the normalized entropy of r_i in the collection $\mathcal{N}$. The global weighting implied by $1 - \varepsilon_i$ reflects the fact that two units appearing with the same count in c_j do not necessarily convey the same amount of information about the composition; this is subordinated to the distribution of the units in the collection $\mathcal{N}$.

If we denote by $\tau_i = \sum_j \kappa_{i,j}$ the total number of times r_i occurs in $\mathcal{N}$, the expression for ε_i is easily seen to be

$$\varepsilon_i = -\frac{1}{\log N} \sum_{j=1}^{N} \frac{\kappa_{i,j}}{\tau_i} \log \frac{\kappa_{i,j}}{\tau_i}. \qquad (2.2)$$

By definition, $0 \leq \varepsilon_i \leq 1$, with equality if and only if $\kappa_{i,j} = \tau_i$ and $\kappa_{i,j} = \tau_i/N$, respectively. A value of ε_i close to 1 indicates a unit distributed across many compositions throughout the collection, while a value of ε_i close to 0 means that the unit is present only in a few specific compositions. The global weight $1 - \varepsilon_i$ is therefore a measure of the indexing power of the unit r_i.

2.2   VECTOR REPRESENTATION

The $(M \times N)$ unit-composition matrix W resulting from the above feature extraction defines two vector representations for the units and the compositions. Each unit r_i can be uniquely associated with a row vector of dimension N, and each composition c_j can be uniquely associated with a column vector of dimension M. Unfortunately, these vector representations are unpractical for three related reasons. First, the dimensions M and N can be extremely large; second, the vectors r_i and c_j are typically very sparse; and third, the two spaces are distinct from one other.

2.2.1   Singular Value Decomposition

To address these issues, it is useful to employ singular value decomposition (SVD), a technique closely related to eigenvector decomposition and factor analysis [52]. We proceed to perform the (order-R) SVD of W as follows:

$$W \approx \hat{W} = U S V^T, \qquad (2.3)$$

where U is the $(M \times R)$ left singular matrix with row vectors u_i $(1 \leq i \leq M)$, S is the $(R \times R)$ diagonal matrix of singular values $s_1 \geq s_2 \geq \cdots \geq s_R > 0$, V is the $(N \times R)$ right singular matrix with row vectors v_j $(1 \leq j \leq N)$, $R < \min(M, N)$ is the order of the decomposition, and, ^{T} denotes matrix transposition.

The closer R is set to $\min(M, N)$, i.e., the nominal full rank value of W, the closer $\hat{W}$ approximates the original matrix W. As R decreases, the fit increasingly degrades. Yet it can be shown (cf. [52]) that the matrix $\hat{W}$ is the best rank-R approximation to W, for any unitarily invariant norm. This entails, for any matrix A of rank R:

$$\min_{\{A : \mathrm{rank}(A) = R\}} \|W - A\| = \|W - \hat{W}\| = s_{R+1}, \qquad (2.4)$$

where $\| \cdot \|$ refers to the L_2 norm, and s_{R+1} is the smallest singular value retained in the order-$(R+1)$ SVD of W. Obviously, $s_{R+1} = 0$ if R is equal to the rank of W.

2.2.2   SVD Properties

The decomposition (2.3) has a number of useful mathematical properties. As is well known, both left and right singular matrices U and V are column-orthonormal, i.e., $U^T U = V^T V = I_R$ (the identity matrix of order R). Thus, the column vectors of U and V, say Θ_k and Ξ_k ($1 \le k \le R$), respectively, each define an orthonormal basis for the vector space of dimension R spanned by the u_i's and v_j's, whereby

$$r_i = \sum_{k=1}^{R} \left\langle r_i, \Xi_k^T \right\rangle \Xi_k^T = \sum_{k=1}^{R} (u_i S)_k \Xi_k^T, \qquad (2.5)$$

$$c_j = \sum_{k=1}^{R} \langle c_j, \Theta_k \rangle \Theta_k = \sum_{k=1}^{R} (v_j S)_k \Theta_k, \qquad (2.6)$$

where $\langle , \rangle$ denotes the dot product, and $(\cdot)_k$ represents the kth element of the R-dimensional vector considered. The vector space spanned by the u_i's and v_j's is referred to as the *LSM space* $\mathcal{L}$. Equations (2.5) and (2.6) are illustrated in Fig. 2.1(a) and 2.1(b), respectively.

With the above definitions for Θ_k and Ξ_k, it is clear that (2.3) can also be written as

$$\hat{W} = \sum_{k=1}^{R} s_k \Theta_k \Xi_k^T, \qquad (2.7)$$

which is known as the rank-R triplet expansion of W. This expression makes more apparent the inherent tradeoff between reconstruction error—minimizing s_{R+1} in (2.4)—and noise suppression—minimizing the ratio between order-R and order-$(R+1)$ traces $\sum_i s_i$. In practice, the dimension R must be set empirically. It is bounded from above by the (unknown) rank of the matrix W, and from below by the amount of distortion tolerable in the decomposition. Values of R in the range $R = 100$–1000 are the most common.

2.3   INTERPRETATION

Equations (2.5) and (2.6), or equivalently Figs. 2.1(a)–(b), amount to representing each unit and each composition as a linear combination of (hidden) abstract concepts, incarnated by the singular vectors Θ_k and Ξ_k ($1 \le k \le R$). This in turn defines a mapping $(\mathcal{M}, \mathcal{N}) \longrightarrow \mathcal{L}$ between the high-dimensional discrete entities $\mathcal{M}, \mathcal{N}$ and the lower dimensional continuous vector space $\mathcal{L}$. In that mapping, each unit $r_i \in \mathcal{M}$ is uniquely mapped onto the *unit vector*

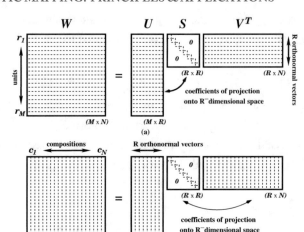

FIGURE 2.1: Illustration of equations (2.5) and (2.6). *(a)* Row vectors (units) are projected onto the orthonormal basis formed by the column vectors Ξ_k $(1 \le k \le R)$ of the right singular matrix V, i.e., the row vectors of V^T. This defines a new representation for the units, in terms of their coordinates in this projection, namely the rows of US: cf. (2.5). Thus, the row vector $u_i S$ characterizes the position of unit r_i in $\mathcal{L}$, for $1 \le i \le M$. *(b)* Similarly, column vectors (compositions) are projected onto the orthonormal basis formed by the column vectors Θ_k $(1 \le k \le R)$ of the left singular matrix U. The coordinates of the compositions in this space are therefore given by the columns of SV^T, or, equivalently, the rows of VS: cf. (2.6). Thus, the row vector $v_j S$ characterizes the position of composition c_j in $\mathcal{L}$, for $1 \le j \le N$.

$\bar{u}_i = u_i S$, and each composition $c_j \in \mathcal{N}$ is uniquely mapped onto the *composition vector* $\bar{v}_j = v_j S$. This dual one-to-one mapping resulting from (2.3) can be illustrated as done in Fig. 2.2.

Since the SVD provides, by definition, a parsimonious description of the linear space spanned by W, the singular vectors are specified to *minimally* span the units in the inventory $\mathcal{M}$ and the compositions in the collection $\mathcal{N}$. Thus, the dual mapping between units/compositions and unit/composition vectors corresponds to an efficient representation of the training data.

A major consequence of this representation is that $\hat{W}$ captures the major structural associations in W and ignores higher order effects. The "closeness" of vectors in $\mathcal{L}$ is therefore determined by the overall pattern of the composition language used in $\mathcal{N}$, as opposed to specific constructs. Hence, two units whose representations are "close" (in some suitable metric) tend to appear in the same kind of compositions, whether or not they actually occur within identical unit contexts in those compositions. Conversely, two compositions whose representations are "close" tend to convey the same meaning, whether or not they contain the same unit constructs.

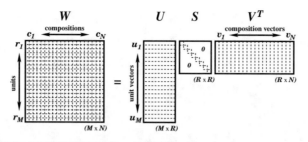

FIGURE 2.2: Singular value decomposition (SVD). Following Figure 2.1, each unit $r_i \in \mathcal{M}$ is uniquely mapped onto the *unit vector* $\bar{u}_i = u_i S$, and each composition $c_j \in \mathcal{N}$ is uniquely mapped onto the *composition vector* $\bar{v}_j = v_j S$.

As a result, units and compositions whose representations are "close" in the LSM space $\mathcal{L}$ tend to be meaningfully related, whether or not such units actually occur in such compositions.

Of course, the optimality of this framework can be debated, since the underlying L_2 norm arising from (2.4) may not the best choice in all situations. We will expand on this comment in Chapter 12. This caveat notwithstanding, the correspondence between closeness in LSM space and meaningful relatedness is well documented, at least when it comes to its LSA incarnation. In applications such as information retrieval, filtering, induction, and visualization, the framework has repeatedly proven remarkably effective in capturing semantic information [26, 58, 64, 69, 85, 102, 103, 140]. The examples presented in Chapter 1 serve to illustrate this observation.

CHAPTER 3

LSM Feature Space

The mapping $(\mathcal{M}, \mathcal{N}) \longrightarrow \mathcal{L}$ automatically resulting from (2.3) opens up the opportunity to apply familiar machine learning techniques in the continuous vector space $\mathcal{L}$. But to do that, it is necessary to first define on $\mathcal{L}$ some suitable metrics consistent with the SVD formalism.

3.1   CLOSENESS MEASURES

Since the matrix W embodies, by construction, all structural associations between units and compositions, it follows that, for a given training collection, $W W^T$ characterizes all co-occurrences between units, and $W^T W$ characterizes all co-occurrences between compositions. Thus, the extent to which units r_i and r_j have a similar pattern of occurrence across the entire collection of compositions can be inferred from the (i, j) cell of $W W^T$; the extent to which compositions c_i and c_j contain a similar pattern of units from the entire inventory can be inferred from the (i, j) cell of $W^T W$; and the extent to which unit r_i globally relates to composition c_j across the entire training collection can be inferred from the (i, j) cell of W itself. This leads to the following comparisons involving units and compositions.

3.1.1   Unit–Unit Comparisons

Expanding $W W^T$ using the SVD expression (2.3), we obtain:[1]

$$W W^T = U S^2 U^T. \tag{3.1}$$

Since S is diagonal, this means that the (i, j) cell of $W W^T$ can be obtained by taking the dot product between the ith and jth rows of the matrix US, namely $u_i S$ and $u_j S$. A natural metric to consider for the "closeness" between units is therefore the dot product (or, equivalently, the cosine of the angle) between $\bar{u}_i$ and $\bar{u}_j$:

$$K(r_i, r_j) = \langle \bar{u}_i, \bar{u}_j \rangle = \cos(u_i S, u_j S) = \frac{u_i S^2 u_j^T}{\|u_i S\| \, \|u_j S\|}, \tag{3.2}$$

---

[1] Henceforth we ignore the distinction between W and $\hat{W}$. From (2.4), this is without loss of generality as long as R approximates the rank of W.

for any $1 \leq i, j \leq M$. A value of $K(r_i, r_j) = 1$ means the two units always occur in the same kind of compositions, while a value of $K(r_i, r_j) < 1$ means the two units are used in increasingly different contexts. While (3.2) does not define a bona fide distance measure in the space $\mathcal{L}$, it easily leads to one. For example, over the interval $[0, \pi]$, the measure

$$\mathcal{D}(r_i, r_j) = \cos^{-1} K(r_i, r_j) \tag{3.3}$$

satisfies the properties of a distance on $\mathcal{L}$ (cf. [10]).

3.1.2   Composition–Composition Comparisons

Proceeding as above, expanding $W^T W$ using (2.3) yields

$$W^T W = V S^2 V^T, \tag{3.4}$$

which means that the (i, j) cell of $W^T W$ can be obtained by taking the dot product between $v_i S$ and $v_j S$. As a result, a natural metric for the "closeness" between compositions is

$$K(c_i, c_j) = \langle \bar{v}_i, \bar{v}_j \rangle = \cos(v_i S, v_j S) = \frac{v_i S^2 v_j^T}{\|v_i S\| \, \|v_j S\|}, \tag{3.5}$$

for any $1 \leq i, j \leq N$. This has the same functional form as (3.2), and therefore the distance (3.3) is equally valid for both unit and composition comparisons.[2]

3.1.3   Unit–Composition Comparisons

Since $W = U S V^T$, the (i, j) cell of W can be obtained by taking the dot product between $u_i S^{1/2}$ and $v_j S^{1/2}$. Thus, a natural metric for the "closeness" between unit r_i and composition c_j is

$$K(r_i, c_j) = \langle \bar{u}_i S^{-1/2}, \bar{v}_j S^{-1/2} \rangle$$

$$= \cos(u_i S^{1/2}, v_j S^{1/2}) = \frac{u_i S v_j^T}{\|u_i S^{1/2}\| \, \|v_j S^{1/2}\|}, \tag{3.6}$$

for any $1 \leq i \leq M$ and $1 \leq j \leq N$. A value of $K(r_i, c_j) = 1$ means that there is a strong relationship between r_i and c_j, while a value of $K(r_i, c_j) < 1$ means that there is increasingly less evidence that they are meaningfully linked across the entire training collection. Interestingly, (3.6) is functionally equivalent to (3.2) and (3.5), but involves scaling by $S^{1/2}$ instead of S. As before, the transformation (3.3) can be used to turn (3.6) into an actual distance measure.

---

[2]In fact, the measure (3.3) is precisely the one used in Fig. 1.1. Thus, the distances on the x-axis of Fig. 1.1 are $\mathcal{D}(c_i, c_j)$ expressed in radians.

3.2   LSM FRAMEWORK EXTENSION

Clearly, the training collection $\mathcal{N}$ will not comprise all conceivable events that can be produced in the composition language. In order to find adequate representations in the space $\mathcal{L}$ for unobserved units and compositions, it is necessary to devise an appropriate extension to the LSM framework. As it turns out, under relatively mild assumptions, finding a representation for new units and compositions in the space $\mathcal{S}$ is straightforward.

Suppose we observe a new composition $\tilde{c}_p$, with $p > N$, where the tilde symbol reflects the fact that the composition was not part of the original collection $\mathcal{N}$. First, we construct a feature vector containing, for each unit in the underlying inventory $\mathcal{M}$, the weighted counts (2.1) with $j = p$. This feature vector $\tilde{c}_p$, a column vector of dimension M, can be thought of as an additional column of the matrix W. Thus, provided the matrices U and S do not change, (2.3) implies

$$\tilde{c}_p = U S \tilde{v}_p^T, \tag{3.7}$$

where the R-dimensional vector $\tilde{v}_p^T$ acts as an additional column of the matrix V^T. This in turn leads to the definition

$$\tilde{v}_p = \tilde{v}_p S = \tilde{c}_p^T U. \tag{3.8}$$

The vector $\tilde{v}_p$, indeed seen to be functionally similar to a composition vector, corresponds to the representation of the new composition in the space $\mathcal{L}$.

To convey the fact that it was not part of the SVD extraction, the new composition $\tilde{c}_p$ is referred to as a *pseudo-composition*, and the new representation $\tilde{v}_p$ is referred to as a *pseudo-composition vector*. Figure 3.1 illustrates the resulting extension to the LSM framework, along with an important caveat further discussed below. Recall that the singular vectors in the SVD expansion (2.3) are specified to minimally span $\mathcal{M}$ and $\mathcal{N}$. As a result, if the new composition contains language patterns which are inconsistent with those extracted from W, (2.3) will no longer apply. Similarly, if the addition of $\tilde{c}_p$ causes the major structural associations in W to shift in some substantial manner, the singular vectors will become inadequate. Then U and S will no longer be valid, in which case it would be necessary to recompute (2.3) to find a proper representation for $\tilde{c}_p$. If, on the other hand, the new composition[3] generally conforms to the rest of the collection $\mathcal{N}$, then $\tilde{v}_p$ in (3.8) will be a reasonable representation for $\tilde{c}_p$.

Once the representation (3.8) is obtained, the "closeness" between the new composition $\tilde{c}_p$ and any other entity in $\mathcal{L}$ can then be computed using any of the metrics (3.2), (3.5), or (3.6) specified above.

---

[3] Any newly observed unit can similarly be treated as an additional row of the matrix W, giving rise to an analogous pseudo-unit vector.

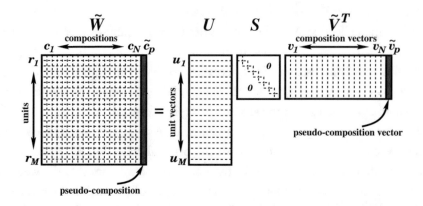

FIGURE 3.1: LSM Framework Extension. *Caveat:* Suppose training is carried out for a banking application involving the word "bank" taken in a financial context. Now suppose $\tilde{c}_p$ is germane to a fishing application, where "bank" refers to the edge of a river or a lake. Clearly, the closeness of "bank" to, e.g., "money" and "account," would become irrelevant. Conversely, adding $\tilde{c}_p$ to W would likely cause such structural associations to shift substantially, and perhaps even disappear altogether. In that case the recomputation of (2.3) would become necessary.

3.3   SALIENT CHARACTERISTICS

Overall the LSM feature space has a number of interesting and fairly generic characteristics. To summarize, the mapping entails:

- a single vector embedding for both units and compositions in the same continuous vector space $\mathcal{L}$;

- a relatively low dimensionality, which makes operations such as clustering meaningful and practical; and

- an underlying topological structure reflecting globally meaningful relationships, with well-motivated, natural similarity metrics to measure the distance between units, between compositions, or between units and compositions in $\mathcal{L}$.

Together these characteristics define the domain of applicability of LSM. Any task where such a representation could be useful could potentially benefit from the mapping. The issue therefore shifts to the definition of units and compositions which is the most appropriate for a given environment.

In natural language applications such as information retrieval, $\mathcal{M}$ is a vocabulary of words, and $\mathcal{N}$ is a text corpus of documents, so units are words, and compositions are documents. The measures (3.2) and (3.5) are useful for (data-driven) semantic classification, such as word/document clustering and related applications (cf. Chapters 6–7), while the measure (3.6) can be exploited for language modeling (cf. Chapter 8).

But the generic nature of the LSM properties also makes the framework applicable to a variety of other tasks in spoken language processing (cf. Chapters 9–11) and possibly beyond (cf. Chapter 12).

CHAPTER 4

Computational Effort

At this point it is legitimate to wonder exactly what the LSM approach, as described in the previous Chapters, entails in terms of computational complexity. We now address this issue for both offline (i.e., computing the SVD) and online (i.e., computing a distance value in the LSM space) calculations.

4.1 OFF–LINE COST

In many applications, W will be a large sparse matrix. In such a case, classical methods for determining the SVD of dense matrices (see, for example, [28]) are clearly not optimal. Because these methods apply orthogonal transformations (Householder or Givens) directly to the input matrix, they incur excessive fill-in and thereby require tremendous amounts of memory. In addition, they compute all the singular values of W; but here very often $R \ll \min(M, N)$, and therefore doing so is computationally wasteful.

Instead, it is more appropriate to solve a sparse symmetric eigenvalue problem, which can then be used to indirectly compute the sparse singular value decomposition. Several suitable iterative algorithms have been proposed by Berry, based on either the subspace iteration or the Lanczos recursion method [25]. The primary cost of these algorithms lies in the total number of sparse matrix–vector multiplications required. Let us denote by Δ_W the density of W, defined as the total number of nonzero entries in W divided by the product of its dimensions, MN. Then the total cost in floating-point operations per iteration is given by [25]

$$\mathcal{N}_{\text{svd}} = 2R[(1 + \Delta_W N)M + (1 + \Delta_W M)N]. \qquad (4.1)$$

In a typical natural language application, such as information retrieval, Δ_W hovers in the range 0.25–0.5% (cf. [64]), and the value of R is between 100 and 250. This expression can therefore be approximated by

$$\mathcal{N}_{\text{svd}} \approx (4R\Delta_W)MN \approx \gamma MN, \qquad (4.2)$$

where γ is a small multiplier, normally satisfying $1 \leq \gamma \leq 5$. For the values of M and N mentioned earlier, $\mathcal{N}_{\text{svd}}$ is thus on the order of 10–50 billion floating-point operations (flops) per iteration. On any midrange desktop machine (such as a 2.66 GHz quad-core Apple Mac Pro, which peaks at approximately 45 Gflops), this translates into maybe up to a few seconds of CPU time. As convergence is typically achieved after 100 or so iterations, the entire decomposition is usually completed within a matter of minutes. Of course, in practice this is still too slow for real time operation, which makes updating the matrix of particular interest, as will be discussed later on in this chapter.

4.2 ONLINE COST

The cost incurred during actual processing stems from two separate contributions:

- the construction of the pseudocomposition representation in $\mathcal{L}$, obtained via (3.8); and
- the computation of one of the metrics (3.2), (3.5), or (3.6).

Although there will often be an additional cost related to integrating the LSM knowledge with some other source of information, this is usually performed independently of the LSM calculations and need not be taken into account here.

The cost of (3.8) would ordinarily be $\mathcal{O}(M)$; however, it is often possible to compute $\tilde{v}_p$ directly in the space $\mathcal{L}$. In [48], for example, this approach results in only $5R + 1$ flops per context instantiation. Similarly, with the proper implementation the cost of computing any of (3.2), (3.5), or (3.6) can be shown to be $R(2R - 1)$ flops per unit or composition [9].

The total online cost to compute an LSM score, per unit and pseudocomposition, is thus obtained as

$$\mathcal{N}_{\text{lsm}} = 2(R + 1)^2 - 1 = \mathcal{O}(R^2). \tag{4.3}$$

For typical values of R, this amounts to less than 0.05 Mflops, which arguably represents a very modest overhead. This bodes well for real-time implementations of distance computations in the LSM space.

4.3 POSSIBLE SHORTCUTS

Still, when the collection $\mathcal{N}$ is very large scale and/or is continually updated, the implementation of LSM is complicated by the computational complexity and storage requirements of the underlying SVD algorithm. This has sparked interest in various shortcuts, often justified by the constraints inherent to the task considered. These shortcuts fall into three categories, depending on whether they result from (i) an alternative implementation of the SVD, (ii) a different matrix decomposition altogether, or (iii) a substantial reformulation of the problem.

4.3.1   Incremental SVD Implementations

A traditional SVD algorithm applies "in-core" computation, i.e., all the matrix components must be loaded into memory before computation can start. As the size of the matrix W grows, the naive approach of using automatic virtual memory support from the underlying operating system may become infeasible. This has led a technique known as "out-of-core" SVD [110], in which data segments (e.g., compositions) are loaded from disk sequentially, and fold-in updates of rows and columns are performed incrementally. Obviously, this fold-in update process entails some degradation in accuracy, mainly due to the loss of orthogonality between columns of U and/or V. However, metrics can be defined to measure this loss, and thereby decide when to replace continuous updates by a classical SVD recomputation. This makes it theoretically possible to consider values of M and N as large as 5,000,000 [110].

Other incremental SVD decomposition algorithms use more sophisticated updates based on the decomposition of the additional data into parallel and orthogonal components [34]. This allows for the handling of arbitrary missing or untrusted values, correlated uncertainty across rows or columns of W, and any available priors. However, in practice, care must still be taken to counter numerical errors that may lead to a loss of orthogonality. This usually requires monitoring the singular vectors as they evolve, and regularly reorthogonalizing the accumulated singular matrices via the application of modified Gram–Schmidt orthogonalization [34].

4.3.2   Other Matrix Decompositions

The SVD is the most common example of a two-sided orthogonal decomposition, where the middle matrix (S) is diagonal. It could be argued, however, that preserving the diagonality of the middle matrix is not essential, as long as its rank-revealing property is maintained. This forms the basis for replacing the SVD with two-sided orthogonal decompositions with (rank-revealing) lower trapezoidal or upper trapezoidal middle matrices, referred to as ULV/URV decompositions [27]. It has been shown that such decompositions produced good approximations to left and right singular subspaces, at a reduced computational cost. Similarly, ULV- or URV-updating compares favorably with the standard fold-in update process.

Another possibility is to replace the SVD by a simplified representation known as the semidiscrete matrix decomposition [100]. In this approach, the basic premise is that the SVD contains more information than is necessary for similarity calculations such as (3.2), (3.5), and (3.6). Instead, the matrices U and V are constrained to only comprise entries taken from the set $\{-1, 0, 1\}$, and a greedy algorithm is used to construct each of the corresponding terms in a triplet expansion similar to (2.7). The semidiscrete decomposition was shown to be a competitive alternative to the SVD for information retrieval tasks, while requiring only a small fraction of the storage space [100]. Note, however, that it is only applicable in situations which call for a similarity score, as opposed to the LSM space itself.

4.3.3   Alternative Formulations

The semidiscrete matrix decomposition can be viewed as a particular implementation of a more general approach known as random mapping [68] or random indexing [93]. The idea behind random mapping is to avoid computing the true (orthogonal) principal components by projecting the data onto a set of nearly orthogonal directions, which then (for all practical purposes) play the role of these principal components. This formulation thus allows for an LSM-like space representation, though compared to LSM, there is an additional knob to tune, the degree of "near orthogonality." Random mapping avoids the computation of the SVD, at the expense of about an order of magnitude larger dimensionality (to compensate for the loss of orthogonality). It also entails a somewhat larger footprint and slower decoding (due to more sizable mappings), and is best applied on corpora of sufficient size, otherwise near orthogonality cannot be achieved [93].

A random mapping is itself an instance of functional graph which tends to have one giant component and a few large trees [68]. More generally, spectral graph theory can be useful for constructing a representation for data sampled from a low-dimensional manifold embedded in a higher dimensional space [24]. Drawing on the correspondence between the graph Laplacian, the Laplace–Beltrami operator on a manifold, and the connections to the heat equation, the algorithm proposed in [24], for example, provides a computationally efficient approach to nonlinear dimensionality reduction that has locality preserving properties and a natural connection to clustering. Compared to LSM, this solution reflects a greater reliance on the intrinsic geometric structure of the underlying manifold [24].

We will close this chapter with an alternative approach based on polynomial filtering [99], which was recently developed to avoid computing the SVD when only a similarity score (3.2), (3.5), or (3.6) is required (as with the semidiscrete matrix decomposition). In this approach, the similarity score (3.2), for example, is computed directly from W by using an appropriate matrix polynomial operating on $W^T W$. This polynomial is chosen to act as a step function whose cutoff is keyed off the singular value s_R. Of course, in practice, the degree of the polynomial is finite, which means the step function cannot be perfectly realized. This therefore entails an extra approximation, whose precise impact is not immediately obvious and needs to be further analyzed [99].

CHAPTER 5

Probabilistic Extensions

One potential shortfall of the LSM approach is that the mapping $(\mathcal{M}, \mathcal{N}) \longrightarrow \mathcal{L}$, being grounded in linear algebra, does not readily lend itself to a probabilistic interpretation. This has sparked interest in various probabilistic extensions of the framework.

5.1 DUAL PROBABILITY MODEL

The first approach is to directly parameterize the components of (2.3) with "reasonable" statistical distributions. This is the case, for example, of the probability model defined in [59] under certain assumptions on the distribution of the input data. In this section, we generalize results reported in [59] to obtain a dual probability representation of LSM.

5.1.1 Composition Model

The basic premise is that the probability of occurrence of a composition c_j in the corpus $\mathcal{N}$ is related to its similarity to a set of canonical, mutually orthogonal "latent structure vectors" z_k, $1 \leq k \leq R$. Furthermore, the compositions are assumed to be identically distributed according to a Gaussian mixture distribution of the form

$$\Pr(c_j | \mathcal{L}) = \prod_{k=1}^{R} \frac{1}{Z(\mathcal{L})} \exp\{\langle c_j, z_k \rangle^2\}, \qquad (5.1)$$

where the conditioning on $\mathcal{L}$ reflects the underlying SVD decomposition (2.3), and $Z(\mathcal{L})$ is an appropriate normalization factor. The next step is to find the z_k's as the optimal parameters for the probability model subject to the constraints:

$$\|z_k\| = 1, \qquad 1 \leq k \leq R. \qquad (5.2)$$

With the additional assumption of statistical independence, this can be solved within the usual maximum likelihood estimation framework, by considering the following log-likelihood function:

$$L(\{z_k\}) = \log \prod_{j=1}^{N} \prod_{k=1}^{R} \Pr(c_j | z_k) \propto \sum_{k=1}^{R} z_k^T W W^T z_k. \qquad (5.3)$$

Note that (5.3) implicitly assumes that the normalization factor is a constant with respect to the latent structure vectors, which is typically justified by the claim that $Z(\mathcal{L})$ is only weakly dependent on the z_k's [59]. Invoking the Rayleigh–Ritz theorem [76], the optimal solution is

$$z_k = \Theta_k, \qquad 1 \leq k \leq R, \tag{5.4}$$

i.e., the optimal latent structure vectors turn out to be the left singular vectors obtained in Chapter 2. In fact, taking (2.6) into account, the probability model for compositions becomes

$$\Pr(c_j | \mathcal{L}) = \prod_{k=1}^{R} \frac{1}{Z(\mathcal{L})} \exp\{\langle c_j, \Theta_k \rangle^2\} = \frac{1}{Z(\mathcal{L})} \exp\{\|v_j S\|^2\}, \tag{5.5}$$

which is closely aligned with the results of Chapter 2. Basically, the probability model unfolds as though the underlying norm had induced an exponential distribution on the space, presumably with pertinent marginality constraints.

5.1.2   Unit Model

The same reasoning can be applied at the unit level. This time we assume that the units are identically distributed according to the Gaussian mixture distribution

$$\Pr(r_i | \mathcal{L}) = \prod_{k=1}^{R} \frac{1}{Z(\mathcal{L})} \exp\{\langle r_i, z_k \rangle^2\}, \tag{5.6}$$

and seek the set of z_k's to maximize the log-likelihood function

$$L(\{z_k\}) = \log \sum_{i=1}^{M} \prod_{k=1}^{R} \Pr(r_i | z_k) \propto \sum_{k=1}^{R} z_k^T W^T W z_k. \tag{5.7}$$

Under similar assumptions, the optimal solution is now

$$z_k = \Xi_k, \qquad 1 \leq k \leq R, \tag{5.8}$$

i.e., the set of right singular vectors obtained in Chapter 2. Taking (2.5) into account, the probability model for units is therefore given by

$$\Pr(r_i | \mathcal{L}) = \prod_{k=1}^{R} \frac{1}{Z(\mathcal{L})} \exp\{\langle r_i, \Xi_k \rangle^2\} = \frac{1}{Z(\mathcal{L})} \exp\{\|u_i S\|^2\}. \tag{5.9}$$

Again, we would have arrived at the same outcome had the norm simply induced an exponential distribution on $\mathcal{L}$.

5.1.3   Comments

The dual probability model derived above has the merit of quantifying the statistical significance of the latent semantic relationships exposed via the SVD decomposition (2.3). In addition, the likelihoods (5.3) and (5.7) can sometimes be useful to suggest an appropriate value for the dimension of the vector space $\mathcal{L}$ [59].

However, this modeling comes at the expense of many restrictive assumptions on the input data, and it is not entirely clear how reasonable some of these assumptions are in practice. It is likely that they may hold better for some collections than others, thereby intrinsically limiting the applicability of this theoretical framework.

5.2   PROBABILISTIC LATENT SEMANTIC ANALYSIS

One way to circumvent such objections is to refrain from imposing any parametric distribution on the unit and composition vectors. Unfortunately, a nonparametric interpretation is not immediately possible, as the component vectors may well contain noninterpretable negative values. This leads to a technique variously known as aspect modeling or probabilistic latent semantic analysis (PLSA) [84].

PLSA avoids the problem of negative values by using nonnegative matrix factorization (rather than SVD) to model each composition as a weighted combination of a set of nonnegative feature vectors [84–86]. In this approach, the decomposition (2.3) is replaced by [84, 117]:

$$W \approx \hat{W} = Q H^T, \qquad (5.10)$$

where Q is $(M \times R)$ with row vectors q_i $(1 \le i \le M)$, H is $(N \times R)$ with row vectors h_j $(1 \le j \le N)$, $R < \min(M, N)$ as previously, and all elements are positive.

This allows an interpretation of (5.10) as a generative statistical latent class model, also known as aspect model, where (i) a composition is selected with probability $\Pr(c_j)$, (ii) a latent class z_k is selected with probability $\Pr(z_k|c_j)$, and (iii) a unit is generated with probability $\Pr(r_i|z_k)$. Under this model, the joint probability between unit r_i and composition c_j can be formulated as

$$\Pr(r_i, c_j) = \sum_{k=1}^{R} \Pr(r_i|z_k)\,\Pr(z_k|c_j)\,\Pr(c_j), \qquad (5.11)$$

which clearly leads to the identification of Q with $\Pr(r_i|z_k)$ and H with $\Pr(z_k|c_j)$ $(1 \le i \le M, 1 \le j \le N, 1 \le k \le R)$.

In addition, this approach lends itself quite well to iterative training using the EM algorithm [6]. Consider the likelihood function

$$L(\{z_k\}) = \log \prod_{i=1}^{M} \prod_{j=1}^{N} \Pr(r_i, c_j)^{\kappa_{i,j}}$$

$$= \sum_{i=1}^{M} \sum_{j=1}^{N} \kappa_{i,j} \log \Pr(r_i, c_j), \tag{5.12}$$

where, as in Chapter 2, $\kappa_{i,j}$ is the count associated with the ith unit and the jth composition, and observe that (5.11), using Bayes' rule, can also be written as

$$\Pr(r_i, c_j) = \sum_{k=1}^{R} \Pr(r_i|z_k) \Pr(c_j|z_k) \Pr(z_k). \tag{5.13}$$

The E-step is then given by

$$\Pr(z_k|r_i, c_j) = \frac{\Pr(r_i|z_k) \Pr(c_j|z_k) \Pr(z_k)}{\displaystyle\sum_{\ell=1}^{R} \Pr(r_i|z_\ell) \Pr(c_j|z_\ell) \Pr(z_\ell)}, \tag{5.14}$$

while the M-step follows from

$$\Pr(r_m|z_k) = \frac{\displaystyle\sum_{j=1}^{N} \kappa_{m,j} \Pr(z_k|r_m, c_j)}{\displaystyle\sum_{i=1}^{M} \sum_{j=1}^{N} \kappa_{i,j} \Pr(z_k|r_i, c_j)}, \tag{5.15}$$

$$\Pr(c_n|z_k) = \frac{\displaystyle\sum_{i=1}^{M} \kappa_{i,n} \Pr(z_k|r_i, c_n)}{\displaystyle\sum_{i=1}^{M} \sum_{j=1}^{N} \kappa_{i,j} \Pr(z_k|r_i, c_j)}, \tag{5.16}$$

$$\Pr(z_k) = \frac{\displaystyle\sum_{m=1}^{M} \sum_{j=1}^{N} \kappa_{i,j} \Pr(z_k|r_i, c_j)}{\displaystyle\sum_{m=1}^{M} \sum_{j=1}^{N} \kappa_{i,j}}. \tag{5.17}$$

In other words, PLSA relies on the likelihood function of multinomial sampling, and aims at an explicit maximization of the predictive power of the model. As is well known [84–86], this corresponds to a minimization of the cross-entropy (or Kullback–Leibler divergence) between the empirical distribution and the model. Thus, PLSA provides a probabilistic framework in terms of a mixture decomposition (the convex combination of the aspects), which has a well-defined probability distribution and can be trained using familiar statistical tools.

5.3   INHERENT LIMITATIONS

On the other hand, PLSA also has a number of inherent limitations. Most critically, (5.11) assumes that, conditioned on the state of every latent class, units and compositions are statistically independent, which in practice is obviously violated for many, if not most, latent classes. As already noted for the dual probability model discussed earlier, this assumption would seem to somewhat restrict the practical validity of the framework.

Also, because each training composition has its own set of latent class parameters, PLSA does not provide a generative framework for describing the probability of an unseen composition [33]. In addition, the number of parameters grows linearly with the size of the collection, which makes the model prone to overfitting, and EM training only converges to a local minimum, so that the quality of the model depends on the initial parameters selected. Finally, likelihood values obtained across different initializations do not correlate well with the accuracy of the corresponding models [35].

Training issues can be addressed in a number of different ways, including the use of multiple random initialization [85], unsupervised clustering [66], and classical LSA [35] to derive a good set of initial parameters. To address the other shortcomings of PLSA, one solution is to rely on latent Dirichlet allocation (LDA). In that approach, a Dirichlet distribution is imposed on the latent class parameters corresponding to compositions in the collection [33]. With the number of model parameters now dependent only on the number of latent classes and unit inventory size, LDA suffers less from overfitting and provides probability estimates of both observed training and unseen test compositions. Further extensions have recently been proposed to the basic LDA framework for specific tasks: see, for example, [126, 145].

We will close this Chapter by mentioning another application of nonnegative matrix factorization, which is more similar in spirit to what will be discussed in Chapter 11. The decomposition (5.10) was recently exploited in [136] for the purpose of automatic auditory scene analysis. In that application, the matrix factorization operates on a magnitude spectrogram of the signal, for the specific purpose of dimensionality reduction. Thus the goal is not so much to formulate the problem in a probabilistic way as it is to expose useful characteristics of the signal. We will come back to this point in Chapter 11.

Part II

Applications

CHAPTER 6

Junk E–Mail Filtering

In applications such as information retrieval, units are words, and compositions are documents. It is in this context that the term "semantic" in LSM comes closest to its ordinary meaning.[1] With units and compositions so defined, the LSM properties can be exploited in several areas of natural language processing. One of the conceptually simplest applications is junk e-mail filtering. To see how the LSM framework can be leveraged in this domain, let us first review some of the conventional approaches to the problem.

6.1 CONVENTIONAL APPROACHES

The goal of junk e-mail filtering is to sort out and discard unsolicited commercial e-mail, commonly known as spam [1, 21, 134]. Traditional methods to cope with a junk e-mail can be roughly classified into three categories: header analysis, rule-based predicates, and/or machine-learning approaches.

6.1.1 Header Analysis

Two early techniques for spam detection and tracking, still widely used in most e-mail servers, are bulk mailing blocks and originating address tracking. The first solution recognizes a message sent in bulk to multiple recipients, and tags it as likely spam. The second approach keeps track of e-mail addresses that have been known to send junk mail, collects them into blacklists, and flags messages originating from them.

The problem is that these simple-minded methods can be fairly easily defeated. To circumvent bulk mailing blocks, some commercial e-mailers routinely rout e-mails through an opaque maze of servers, so the message ends up not looking like a bulk message; in fact, they often rely on sophisticated scanners to locate insecure networks of unsuspecting third parties through which messages can be relayed. To render tracking ineffective, they change their originating address often, and liberally, jumping from one of a large block of sender addresses to the next when complaints flow. They also routinely use specialized software to

---

[1]Recall, however, that this terminology still has a rather narrow interpretation in terms of words taken as referents to the general topic of the document.

replace the originator's identification in e-mail headings with a false one. This has spawned a new cottage industry peddling "spamware," or high-tech tools needed by spammers, such as cloaking technology and stealth e-mailing products. One such company even throws in 25 million free "fresh e-mail addresses" with any bulk e-mail software purchase.

6.1.2   Rule-Based Predicates

Rule-based approaches (cf., for example, [49]) are more immune to all this "cloaking," since they do not depend on sender information. They generally scan the subject and/or the body of the e-mail message for some predetermined keywords, or strings of keywords (keyphrases), or specific layout features, the presence of which tags the message as suspicious.

However, spammers have also become quite adept at complicating the rules. To defeat those looking only at the subject field, they come up with creative subjects (e.g., "In response to your query") in order to compel the recipient to view the message. To defeat those looking at the entire body of the e-mail, they carefully phrase their message to avoid predictable keywords. Besides, keyword spotting is not without problems of its own. Consider an attempt to detect pornographic material by using the keyword "sex:" not only would this approach still let through a great deal of "erotica," but it might also block out legitimate biology articles discussing "sex differences," not to mention any literary novel with a passing reference to the "fair sex." As a result, typical error rates are often as high as 20%.

To determine whether the general topic of the message is consistent with the user's interests, one possibility is to characterize this topic in terms of rules, and to exploit rule-based text classifiers. In [139], for example, semantic and syntactic textual classification features are used for the detection of abusive newsgroup messages through a probabilistic decision tree framework. Unfortunately, this approach suffers various practical deficiencies. First, the user has to handcraft a set of rules appropriate to his/her own usage, a complex endeavor even for a highly experienced and knowledgeable user. Second, the rules have to be continuously updated to prevent any work-arounds from mass e-mail senders, a tedious task to carry out manually. And finally, such rules typically result in a binary classification, which gives no information regarding the degree of relevance to the user's interests.

6.1.3   Machine Learning Approaches

Such issues have sparked interest in various ways to automatically learn, modify, and evaluate the relevant parameters. Memory-based learning, for example, has been applied to a suitable vector encoding of the rules [53]. Bayesian classifiers [124], boosting trees [40], support vector machines [62], and combinations thereof [132], are other machine-learning techniques that have been applied to text categorization in general and e-mail filtering in particular.

Among these, Bayesian methods are particularly attractive, because they more formally model the relationship between the content of a message and the informational goals of the user [104]. Bayesian classifiers are based on a probabilistic model of text generation, where first a class is chosen according to some prior probability, and then a text is generated according to a class-specific distribution. Thus the classifier always outputs the class which is most likely to have generated the document. This approach has proven effective when used with suitable features [131]. Basically, the idea is to represent each message by a vector of carefully selected attributes, and to evaluate this vector in the underlying vector space. If the vector falls in a region largely populated by spam messages, it is flagged as suspicious.

Since the model parameters are estimated from training examples that have been annotated with their correct class, the attribute vector is by nature tied to the underlying classes. Following typical text categorization solutions (deployed in such domains as news [80] and mail [121]), Bayesian attributes are selected on the basis of mutual information, with the relevant probabilities estimated as frequency ratios. This, however, tends to result in an uneven handling of synonyms and an inherent discounting of rare words.

In addition, to enable estimation in the first place, all attributes are assumed to be conditionally independent given the category: this is usually referred to as the "Naive Bayes" solution. Clearly, this assumption is violated in most real world situations, leading to a degradation in classification accuracy due to "information double-counting" and interaction omission. In practice, performance tends to deteriorate when the number of attributes grows beyond a few hundred. As more attributes are usually required in order to capture the variability of (potentially multilingual) messages, modifications become necessary for handling the conditional dependence, and confer to this approach the wider applicability it needs for the general detection of a junk e-mail. Such modifications, which include feature selection, feature grouping, and explicit correlation modeling, have become a lively research area (cf., e.g., [105]).

6.2   LSM-BASED FILTERING

LSM-based filtering adopts a completely different point of view, centered around whether or not the latent subject matter is consistent with the user's interests. So far the topic of a message has been characterized more or less explicitly, using either specific keywords/rules or the selection of specific attributes.[2] Alternatively, the topic of a message can be defined as an emergent entity loosely characterized by the particular pattern of word co-occurrences observed in the e-mail.

---

[2]This explains, in particular, why rare words and synonyms may not be handled optimally: in the case of attributes, the relevant mutual information is likely to be inaccurately estimated, and in the case of rules, such words are unlikely to surface at all. Yet, this information is often key to the determination of a particular topic, so it needs to be fully exploited.

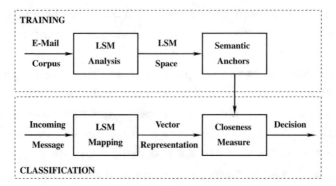

FIGURE 6.1: LSM–based junk e–mail filtering. During training, LSM is performed on a suitable e-mail corpus, typically assembled from a (limited) number of previously received legitimate and junk messages. The latent semantic information thus extracted characterizes what constitutes legitimate versus junk e-mail. This knowledge is automatically embedded in a vector space featuring two semantic anchors, one representing the centroid of all e-mails that are of interest to the user, and the other representing the centroid of all e-mails that are not. The centroid of each cluster is the semantic representation of the associated category in the LSM space. Then, each incoming message is evaluated against these two anchors using the closeness measure (3.5), which indicates how related it is to what the user likes and dislikes.

This is where LSM comes in. Since the goal is to analyze the content of the message to determine whether it is spam or legitimate, the problem can be viewed as a case of document clustering into only two semantic categories: legitimate and junk. The idea is therefore to represent these categories as two *semantic anchors* in the LSM space. The implementation follows the two phases (training and classification) depicted in Fig. 6.1.

In this approach, $\mathcal{M}$ is an inventory of words, symbols, and combination thereof, and $\mathcal{N}$ is a (binary) collection of legitimate and junk e-mail messages. Two semantic anchors are established in the space $\mathcal{L}$, one for each category. During training, we decompose the matrix of co-occurrences as shown in Fig. 6.2, which is readily recognized as the rank-2 version of Fig. 2.2. The mapping here is between (i) the aggregate collection of legitimate e-mail messages and the two-dimensional vector $\bar{v}_1 = v_1 S$, and (ii) the aggregate collection of unsolicited e-mail messages and the two-dimensional vector $\bar{v}_2 = v_2 S$.

During classification, each incoming e-mail message is mapped onto $\mathcal{L}$ using (3.8), as illustrated in Fig. 6.3. Again, this is but a version of Fig. 3.1. The resulting message vector is classified against the two anchors. If it is deemed "closer," in the sense of (3.5), to the anchor representing a legitimate e-mail, it is let through. Otherwise, it is tagged as junk, and optionally discarded, if confidence is high enough. The procedure is completely automatic and requires

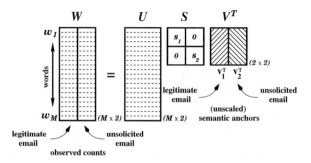

FIGURE 6.2: Finding the semantic anchors. Recall from the semantic nature of the mapping that the relative positions of the anchors are determined by the overall pattern of the language used in $\mathcal{T}$, as opposed to specific keywords or constructs. Hence, a vector "close"—in the metric (3.5)—to $\bar{v}_1$ can be expected to be associated with a legitimate e-mail, while a vector "close" to $\bar{v}_2$ will tend to be associated with an unsolicited e-mail. Thus, classification can be based on the entire content of each message, as opposed to the presence or absence of a small number of keywords.

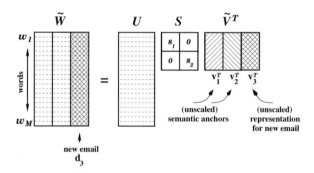

FIGURE 6.3: Vector representation for a new e–mail. Recall from (3) that, technically speaking, the representation of the message vector is only an approximation, since the new e-mail was not part of the SVD extraction. If the incoming message contains language patterns which are inconsistent with those extracted from the original W, the SVD expansion will no longer apply. Similarly, if the addition of the new message causes the major structural associations in W to shift in some substantial manner, U and S will no longer be valid. If, on the other hand, the new e-mail generally conforms to the rest of the corpus $\mathcal{T}$, then $\bar{v}_3$ will be a reasonable representation.

no input from the user. In case the user chooses to correct misclassifications, however, the framework can also be adapted to learn new junk topics, and the semantic anchors updated accordingly, for a more robust outcome.

6.3   PERFORMANCE

The performance of LSM-based filtering was investigated in detail in [21], and was found to be competitive with state-of-the-art spam filtering methods based on machine-learning [1, 134]. We performed experiments on a publicly available e-mail corpus (*Ling-Spam* [1]), which consists of 2412 legitimate and 481 spam messages, divided into ten parts of equal size (with equal proportion of legitimate and spam message across the ten parts). For ease of comparison, we adopted exactly the methodology followed in [134]: we used ten-fold crossvalidation in the experiments (using nine parts for training and the remaining part for testing) with a different test set in each trial, and averaged the evaluation measure across the ten iterations. All experimental conditions were identical to [134].

The results are summarized in Table 6.1, where P_{cor} refers to the probability of correct classification. Interestingly, the latent semantic solution seems to be less biased toward one class, while still achieving an overall accuracy competitive with the state-of-the-art. Note that in this corpus spam messages represent only 16.6% of the data. Had the database been balanced, the above figures would entail an overall accuracy of 94.22%, 97.91%, and 98.34% for the two Naive Bayes algorithms and the latent semantic approach, respectively. The parsimonious nature of LSM therefore seems to contribute to a more robust generalization from the seen to the unseen. The resulting system has been a part of the e-mail client bundled with MacOS X since August 2002.

We conclude this Chapter by underscoring the dynamic nature of the field. Spammers are constantly dreaming up new, creative ways to defeat junk e-mail filters, and the

TABLE 6.1: Comparison with Naive Bayes Approaches. "Baseline NB" refers to the standard Naive Bayes algorithm using a multi-variate Bernoulli statistical event model, "Advanced NB" to a more sophisticated Naive Bayes approach using a multinomial event model, and "Latent Semantic" to the method proposed above. Figures for the two Naive Bayes approaches are those from [134]. The "Overall" column reflects the weighted performance on both legitimate and spam messages.

APPROACH CONSIDERED	LEGIT P_{cor}	JUNK P_{cor}	OVERALL P_{cor}
Baseline NB	98.88%	88.57%	98.00%
Advanced NB	99.34%	96.47%	98.86%
Latent semantic	98.76%	97.92%	98.62%

classification methodology adopted (such as LSM versus Naive Bayes algorithms) is only one aspect of the "war" against spam. In recent years, spammers have turned to increasingly sophisticated obfuscation techniques, including character string modifications (such as replacing "a" by "@", inserting blank spaces or other distracting symbols, etc.), additions of unrelated, legitimate-looking literary material (as header or footer), image wrappers, and short messages which are difficult to parse properly (such as a single URL). Such obfuscation attempts makes message preprocessing (tokenization, stripping, image handling, etc.) an ever more important component of any junk e-mail filter.

CHAPTER 7

Semantic Classification

(Data-driven) semantic classification refers to the task of determining, for a given (newly observed) document, which one of the several predefined topics the document is most closely aligned with. Each topic is typically specified during training via a cluster of documents from the collection $\mathcal{N}$. This is clearly in the same vein as the junk e-mail filtering problem covered in Chapter 6: the only difference is that general semantic classification usually involves more than two predefined topics.

7.1 UNDERLYING ISSUES

Semantic clustering is required for many tasks, such as automated call routing [39, 77], desktop interface control [22], and unsupervised tutoring [79]. It is also an integral, if often implicit, part of spoken language understanding [125, 149] and audio-content indexing [108].

Conventional approaches to the problem normally involve HMMs (cf., e.g., [81]) or neural networks (cf., e.g., [98]). In both cases, there is a strong reliance on local context, and therefore classification performance tends to be tied to the particular word sequences observed, rather than the broad concepts expressed in the documents. Not only is it far from ideal for *semantic* classification, but in practical terms, the main drawback of this framework is the lack of flexibility inherent in the dual goal of the language model, which has to be specific enough to yield semantically homogeneous clusters, and yet inclusive enough to allow for proper generalization.

7.1.1 Case Study: Desktop Interface Control

To illustrate the point, consider the particular task of desktop interface control. This is an example of spoken language processing involving "command and control" speech recognition, for the purpose of controlling a variety of user interface elements by voice [22]. Because this task is restricted to a well-defined domain of discourse, the natural choice for the underlying language model is a formal grammar,[1] which tightly constrains the search for the most likely

---

[1]By formal grammar, we mean a syntactic description of composite expressions for the domain of interest, as exemplified by a context-free grammar. In practice, it is expedient to invoke a regularity assumption in order to

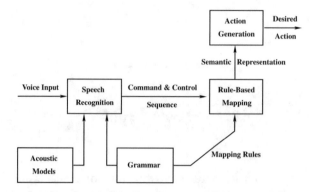

FIGURE 7.1: Traditional desktop interface control. The typical system comprises three major components: a speech recognition engine, a mapping module, and an action generation engine. The role of speech recognition is to determine which command/query sequence (among those supported by the grammar) the user's voice input corresponds to. For example, the user might say: "*what's the time?*," which the recognition engine might recognize as the supported sequence "*what is the time?*". This sequence is then transformed into a semantic representation through some task-specific mapping rules. Obviously, these rules are closely tied to what can be parsed by the grammar. Finally, the semantic representation is converted to the desired action via the action generation engine. In the example above, that would be the appropriate operating system call to query the time.

sequence of words uttered [92]. This in turn has a dual benefit. First, it improves recognition performance by eliminating hypotheses which fall outside the grammar. Second, it simplifies the translation from recognized utterance to action to be taken, by establishing a priori an unambiguous mapping between parses and actions.

The resulting framework is depicted in Fig. 7.1. Many command and control systems conforming to this framework have reached commercial deployment, and routinely operate in speaker-independent mode under adverse environmental conditions (e.g., over the telephone, in conjunction with a far-field microphone, or in low signal-to-noise ratio situations) [101, 123].

7.1.2   Language Modeling Constraints

Close examination of the role of grammar in Fig. 7.1 reveals an attempt to enforce two—largely contradictory—sets of constraints:

- language constraints, to keep the size of the search space under control, so that recognition can be acceptably fast and accurate, and

leverage input string matching on a finite-state automaton. The outcome, of course, is a finite-state grammar. For our purpose here, the distinction between the two is not essential, so from this point onward we simply use the term "grammar."

- semantic constraints, to gather all wordings which convey the same meaning for each action.

As a result, the grammar is both too impoverished and too rich: too impoverished, because for all but the simplest domains it is not possible to exhaustively cover the task, hence many semantically correct paths may never even enter the search; and yet, too rich, because some of the paths that do may be too idiosyncratic to warrant the extra complexity.

Even more importantly from the point of view of the user, this framework implicitly assumes that one can remember which paths are covered, and which are not. In situations where there may be hundreds of commands, this is not very realistic.

7.2   SEMANTIC INFERENCE

This lack of flexibility in the language model can be alleviated by considering an LSM-based approach. Indeed, LSM disregards word order, and relies only on global evidence. It should therefore lead to a better generalization when it comes to semantic constraints. This assumes that language constraints are enforced via another mechanism, such as an n-gram language model.

7.2.1   Framework

The resulting framework is depicted in Fig. 7.2. Following e-mail filtering, each topic is specified during training via a cluster of documents from the collection $\mathcal{N}$, and uniquely associated with a particular outcome in the task. The corresponding semantic anchors are then calculated in the LSM space. Such semantic anchors are derived without regard to the syntax used to express the semantic link between various word sequences and the corresponding outcome. This opens up the possibility of mapping a newly observed word sequence onto an outcome by computing the distance (3.3) between that document and each semantic anchor, and picking the minimum.

In the context of desktop interface control, this approach is referred to as *semantic inference* [22]. In contrast with usual inference engines (cf. [57]), semantic inference thus defined does not rely on formal behavioral principles extracted from a knowledge base. Instead, the domain knowledge is automatically encapsulated in the LSM space in a data-driven fashion. It relaxes some of the interaction constraints typically attached to the domain, and thus allows the end user more flexibility in expressing the desired command/query.

7.2.2   Illustration

A simple example of semantic inference is presented in Fig. 7.3. It illustrates the main characteristics of the approach: not only is there no a priori constraint on what the user can say, but it can generalize well from a tractably small amount of training data. For example, it can learn

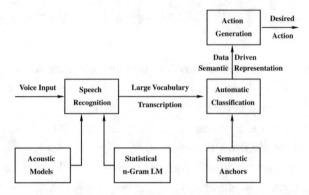

FIGURE 7.2: LSM-based desktop interface control. Now the inference component operates as the back-end of a large vocabulary transcription system. To make this possible, the grammar is replaced by a looser statistical *n*-gram LM, suitably trained for the domain considered. The role of speech recognition is now to transcribe, hopefully with a reasonably low word error rate, the user's voice input into a sequence of words. In principle, there is no restriction on the formulation: the user can express the command in his or her own words. Continuing with the previous example, s/he might choose to say, for instance: "*how late is it?*,"which the transcription engine might now recognize as "*how late is this?*". Accordingly, rule-based mapping is replaced by automatic semantic classification, whose role is to extract the data-driven semantic representation closest in meaning to the user's utterance. The intended action is then generated as previously. In the example, that would happen if (despite the substitution error) "*how late is this?*" is close enough to "*what is the time?*" in the appropriate semantic space.

synonyms and apply them in new contexts. Suppose for instance that the training data contained "*cancel the meeting*" and "*undo the meeting*" as variants of one command, but only "*cancel the last command*" as a variant of another command. Because the variants for the former command indicate that "*cancel*" and "*undo*" are synonyms, the new variant "*undo the last command*" would still be correctly mapped. This in turn tends to reduce the associated cognitive load and thereby enhance user satisfaction.

7.3   CAVEATS

The "bag-of-word" model inherent to LSM is advantageous in applications where local constraints are discarded anyway (such as in information retrieval [26, 58, 64]), or for tasks such as call routing, where only the broad topic of a message is to be identified [39, 47, 74]. For general spoken interaction tasks, however, this limitation may sometimes be more deleterious.

Imagine two desktop interface commands that differ only in the presence of the word "*not*" in a crucial place. The respective vector representations could conceivably be relatively close in the LSM space, and yet have vastly different intended consequences. Worse yet, some

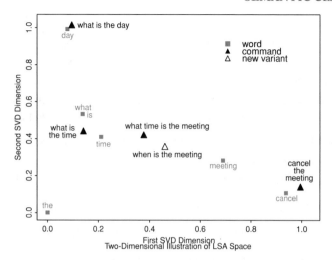

FIGURE 7.3: Semantic Inference in Latent Semantic Space $\mathcal{L}$ ($R = 2$). Consider a task with $N = 4$ outcomes, each represented by a unique command: (i) "*what is the time,*" (ii) "*what is the day,*" (iii) "*what time is the meeting,*" and (iv) "*cancel the meeting.*" There are only $M = 7$ words in the vocabulary, "*what*" and "*is*" always co-occur, "*the*" appears in all four commands, only (ii) and (iv) contain a unique word, and (i) is a proper subset of (iii). Constructing the (7×4) word-command matrix and performing the SVD, we obtain the following latent semantic space $\mathcal{L}$.

The two words which each uniquely identify a command—"*day*" for (ii) and "*cancel*" for (iv)—each have a high coordinate on a different axis. Conversely, the word "*the,*" which conveys no information about the identity of a command, is located at the origin. The semantic anchors for (ii) and (iv) fall "close" to the words which predict them best—"*day*" and "*cancel*", respectively. Similarly, the semantic anchors for (i) and (iii) fall in the vicinity of their meaningful components—"*what–is*" and "*time*" for (i) and "*time*" and "*meeting*" for (iii)—with the word "*time,*" which occurs in both, indeed appearing "close" to both.

Consider now a variant of command (iii) unobserved during training, such as "*when is the meeting,*" which maps onto the hollow triangle in $\mathcal{L}$. As expected, this point is closest to the anchor of command (iii): the variant is deemed semantically most related to (iii), and the correct outcome selected accordingly. In essence, the system has performed elementary, "bottom-up" natural language understanding by inferring that "*when*" is a synonym for "*what time.*"

commands may differ only through word order. Consider the example:

$$\begin{aligned} change\ popup\ to\ window \\ change\ window\ to\ popup \end{aligned} \qquad (7.1)$$

Following conventional LSM, these two commands are mapped onto the *exact same point* in LSM space. This makes them obviously impossible to disambiguate.

We address this issue by extending the basic LSM framework via word agglomeration. The idea is to move from words and commands to word n-tuples and n-tuple commands, where each word n-tuple is the agglomeration of n successive words, and each (n-tuple) command is now expressed in terms of all the word n-tuples it contains. We then construct the associated "word-command" matrix to capture co-occurrences between word n-tuples and n-tuple commands. The rest of LSM proceeds as before, with the LSM space now comprising both word n-tuple vectors and n-tuple command vectors. Clearly, this model incorporates local constraints of span less than or equal to n.

To illustrate, in the simple case $n = 2$, the commands in (7.1) now become, respectively,

$$\begin{aligned} &\textit{\#:change change:popup popup:to to:window window:\#} \\ &\textit{\#:change change:window window:to to:popup popup:\#} \end{aligned} \qquad (7.2)$$

where the symbol # represents a sentence boundary marker, and : is used to separate the two components of each 2-tuple. The effect of this agglomeration is obvious. Whereas at the word level the two commands (7.1) totally overlapped, at the 2-tuple level they only have the rather generic "#:change" in common, a fairly minimal overlap. It has thus become possible to distinguish between the two in the new LSM space.

There are clearly many possible variations on word agglomeration, depending on the maximum choice of n, whether any thresholding or backoff strategy is applied, whether several values of n are entertained simultaneously, etc. The implementation of word agglomeration is therefore heavily dependent on the application considered. As a guideline, note that it does not impact the command dimension of the co-occurrence matrix, but it does increase the word dimension, by a factor theoretically up to M^{n-1}. The (maximum) choice of n therefore centers around the following tradeoff: large enough to encapsulate meaningful local constraints, and small enough to preserve generalization properties and keep the problem computationally tractable. In practice, the familiar choice $n \leq 3$ is compatible with typical instances of spoken interaction.

Semantic inference with word agglomeration was evaluated in [22] on a desktop interface control task involving 113 distinct actions. Under realistic usage conditions, the method exhibited a 2–5% classification error rate. Various training scenarios of increasing scope were also considered to assess the influence of coverage on performance. Sufficient semantic knowledge about the task domain was found to be captured at a level of coverage as low as 70%. This illustrates the good generalization properties of the LSM-based solution.

We conclude this Chapter with a final caveat. Word agglomeration allows semantic inference to handle a modicum of logical structure, such as negation. A predicate-argument structure, however, cannot generally be recovered. In a situation where the predicate-argument structure can be expanded (e.g., by enumeration) into a reasonably sized set of sentences, the method can—implicitly—cope with it.[2] But a command like "*pay* $34.82 *to Pacific Bell*" clearly requires a much more systematic treatment of parameterization. In addition, semantic inference requires a sufficient amount of training data to achieve an acceptable level of performance. Otherwise, the LSM space may fail to capture relevant semantic relationships for the domain of interest. In particular, complete action coverage is required to avoid systematic errors, which in practice requires recomputing the LSM space each time a new action is added.

---

[2]For example, desktop interface control tasks tend to comprise sets of commands that can be parameterized, like: "*sort window by <manner>*," where "*<manner>*" might include *label, size, date modified*, etc. In the evaluation task considered in [22], there were seven possible arguments. Although admittedly more prone to errors than others, these commands were nevertheless handled with a reasonable degree of success.

CHAPTER 8

Language Modeling

Language modeling plays a pivotal role in several areas of language processing. In automatic speech recognition (ASR), for example, it is variously used to constrain the acoustic analysis, guide the search through various (partial) text hypotheses, and/or contribute to the determination of the final transcription [4, 89, 123]. Fundamentally, its function is to encapsulate as much as possible of the syntactic, semantic, and pragmatic characteristics for the task considered. Semantic characteristics, in particular, can be accounted for by the LSM paradigm, which then advantageously complements the standard n-gram approach.

8.1 N-GRAM LIMITATIONS

The purpose of a statistical language model is to compute the probability of the current word, say r_q, based on the admissible history (context) available, H_{q-1} [130]. An important special case is the n-gram model, where the history is taken to be the previous $n-1$ words, i.e., the string $H_{q-1}^{(n)} = r_{q-1} r_{q-2} \cdots r_{q-n+1}$.

In the past two decades, n-gram models have steadily emerged as the preferred way to impose a collection of constraints on word sequences in a wide range of domains [130]. This success notwithstanding, parameter reliability demands low values of n (see, for example, [90, 97, 115, 116]), which in turn imposes an artificially local horizon to the language model. As a result, n-grams as typically derived are inherently unable to capture large-span relationships in the language.

Consider, for instance, predicting the word "*fell*" from the word "*stocks*" in the two equivalent phrases:

$$\textit{stocks fell sharply as a result of the announcement,} \qquad (8.1)$$

and

$$\textit{stocks, as a result of the announcement, sharply fell.} \qquad (8.2)$$

In (8.1), the prediction can be done with the help of a bigram language model ($n = 2$). This is straightforward with the kind of resources which have been available for the past 15 years [127]. In (8.2), however, the value $n = 9$ would be necessary, a rather unrealistic proposition at the present time. In large part because of this inability to reliably capture large-span behavior, the performance of conventional *n*-gram technology has essentially reached a plateau [130].

This observation has sparked interest in a variety of research directions, mostly relying on either *information aggregation* or *span extension* [11]. Information aggregation increases the reliability of the parameter estimation by taking advantage of exemplars of other words that behave "like" this word in the particular context considered. The tradeoff, typically, is higher robustness at the expense of a loss in resolution.

Span extension extends and/or complements the *n*-gram paradigm with information extracted from large-span units (i.e., comprising a large number of words). The tradeoff here is in the choice of units considered, which has a direct effect on the type of long distance dependences modeled. These units tend to be either syntactic (cf. [44, 45, 91, 92, 122, 153]) or semantic (cf. [10, 48, 67, 88, 106, 129, 135]) in nature. Among the latter are those derived from the LSM framework.

8.2   MULTISPAN LANGUAGE MODELING

In LSM language modeling, the history is the current document up to word r_{q-1}, $H_{q-1}^{(l)} = \tilde{c}_{q-1}$, where the (pseudo-)document $\tilde{c}_{q-1}$ is continually updated as q increases. Assuming the document being processed is semantically homogeneous, eventually we can expect the associated pseudo-document vector, which follows from (3.8), to settle down in the vicinity of the document cluster corresponding to the closest semantic content. Thus, the LSM framework can account for the global constraints present in the language, while the *n*-gram approach takes care of local constraints. This forms the basis for the complementarity between the two models [10].

8.2.1   Hybrid Formulation

The next step is to combine the two together. While strategies such as linear interpolation have been considered (cf. [48]), a more tightly integrated formalism is desirable for the overall language model probability. Under relatively mild conditions, it is in fact possible to derive an integrated *n*-gram+LSM formulation. Starting with the definition

$$\Pr(r_q | H_{q-1}) = \Pr(r_q | H_{q-1}^{(n)}, H_{q-1}^{(l)}), \tag{8.3}$$

we make the assumption that the probability of the document history given the current word is not affected by the immediate context preceding it. This reflects the fact that, for a given word, different syntactic constructs (immediate context) can be used to carry the same meaning (document history). This is obviously reasonable for content words. How much it matters for function words is less clear, but we conjecture that if the document history is long enough, the semantic anchoring is sufficiently strong for the assumption to hold.

With this assumption, (8.3) can be written as (cf. [10])

$$\Pr(r_q|H_{q-1}) = \frac{\Pr(r_q|r_{q-1}r_{q-2}\cdots r_{q-n+1})\Pr(\tilde{c}_{q-1}|r_q)}{\sum_{r_i \in \mathcal{M}}\Pr(r_i|r_{q-1}r_{q-2}\cdots r_{q-n+1})\Pr(\tilde{c}_{q-1}|r_i)}. \qquad (8.4)$$

If $\Pr(\tilde{c}_{q-1}|r_q)$ is viewed as a prior probability on the current document history, then (8.4) simply translates the classical Bayesian estimation of the n-gram (local) probability using a prior distribution obtained from (global) LSM. The end result, in effect, is a modified n-gram language model incorporating large-span semantic information: this approach is referred to as *multispan* language modeling.

Note that (8.4) can be further simplified using Bayes' rule. Since the quantity $\Pr(\tilde{c}_{q-1})$ vanishes from both numerator and denominator, we are left with

$$\Pr(r_q|H_{q-1}) = \frac{\Pr(r_q|r_{q-1}r_{q-2}\cdots r_{q-n+1})\frac{\Pr(r_q|\tilde{c}_{q-1})}{\Pr(r_q)}}{\sum_{r_i \in \mathcal{M}}\Pr(r_i|r_{q-1}r_{q-2}\cdots r_{q-n+1})\frac{\Pr(r_i|\tilde{c}_{q-1})}{\Pr(r_i)}}, \qquad (8.5)$$

where $\Pr(r_q)$ is simply the standard unigram probability. Intuitively, the expression $\Pr(r_q|\tilde{c}_{q-1})$ reflects the "relevance" of word r_q to the admissible history, as observed through $\tilde{c}_{q-1}$. As such, it will be highest for words whose meaning aligns most closely with the semantic fabric of $\tilde{c}_{q-1}$ (i.e., relevant "content" words), and lowest for words which do not convey any particular information about this fabric (e.g., "function" words like "*the*"). This behavior is exactly the opposite of that observed with the conventional n-gram formalism, which tends to assign higher probabilities to (frequent) function words than to (rarer) content words. Hence the attractive synergy between the two paradigms.

8.2.2   Context Scope Selection

In practice, expressions like (8.4) are often slightly modified so that a relative weight can be placed on each contribution (here, the n-gram and LSM probabilities). Usually, this is done via empirically determined weighting coefficients. In the present case, such weighting is motivated by the fact that in (8.4) the "prior" probability $\Pr(\tilde{c}_{q-1}|r_q)$ could change substantially as the current document unfolds. Thus, rather than using arbitrary weights, an alternative approach is to dynamically tailor the document history $\tilde{c}_{q-1}$ so that the n-gram and LSM contributions remain empirically balanced.

This approach, referred to as context scope selection, is more closely aligned with the LSM framework, because of the underlying change in behavior between training and recognition. During training, the scope is fixed to be the current document. During recognition, however, the concept of "current document" is ill-defined, because (i) its length grows with each new word, and (ii) it is not necessarily clear at which point completion occurs. As a result, a decision has to be made regarding what to consider "current," versus what to consider part of an earlier (presumably less relevant) document.

A straightforward solution is to limit the size of the history considered, so as to avoid relying on old, possibly obsolete fragments to construct the current context. Alternatively, to avoid making a hard decision on the size of the caching window, it is possible to assume an exponential decay in the relevance of the context [9]. In this solution, exponential forgetting is used to progressively discount older utterances.

8.2.3   LSM Probability

Once this is done, the expression $\Pr(\tilde{c}_{q-1}|r_q)$ can be easily derived from the "closeness" between the associated word vector and pseudodocument vector in $\mathcal{L}$. However, to express the outcome as a probability, it is necessary to go from the distance measure (3.3) to an actual probability measure.

To do that, it is possible to follow, for example, the technique described in Section 5.1, but, as noted there, this has the drawback of placing a number of assumptions on the input data. In practice, it is not necessary to incur this restriction. Considering that $\tilde{c}_{q-1}$ is only a partial document anyway, exactly what kind of distribution is induced is probably less consequential than ensuring that the pseudodocument is properly scoped (cf. the above discussion). Basically, all that is needed is a "reasonable" probability distribution to act as a proxy for the true (unknown) measure.

We therefore opt to use the empirical multivariate distribution constructed by allocating the total probability mass in proportion to the distances observed during training. In essence, this reduces the complexity to a simple histogram normalization, at the expense of introducing a potential "quantization-like" error. Of course, such error can be minimized through a variety of histogram smoothing techniques. Also note that the dynamic range of the distribution typically needs to be controlled by a parameter that is optimized empirically, e.g., by an exponent on the distance term, as carried out in [48].

8.3   SMOOTHING

Since the derivation of (8.5) does not depend on a particular form of the LSM probability, it is possible to take advantage of the additional layer(s) of knowledge uncovered through word and/or document clustering. Basically, we can expect words/documents related to the current document to contribute with more synergy, and unrelated words/documents to be better discounted (cf. [128]). In other words, clustering provides a convenient smoothing mechanism in the LSM space [9].

8.3.1 Word Smoothing

To illustrate, assuming a set of word clusters C_k, $1 \leq k \leq K$, is available, the second term in the numerator of (8.5) can be expanded as follows:

$$\Pr(r_q | \tilde{c}_{q-1}) = \sum_{k=1}^{K} \Pr(r_q | C_k) \Pr(C_k | \tilde{c}_{q-1}). \qquad (8.6)$$

In (8.6), the probability $\Pr(C_k | \tilde{c}_{q-1})$ stems from a unit-composition comparison and can therefore be obtained with the help of (3.6), by simply replacing the representation of the word r_q by that of the centroid of word cluster C_k. In contrast, the probability $\Pr(r_q | C_k)$ depends on the "closeness" of r_q relative to this (word) centroid. To derive it, we therefore have to rely on the empirical multivariate distribution induced not by the distance obtained from (3.6), but by that obtained from the measure (3.2). Note that a distinct distribution can be inferred on each of the clusters C_k, thus allowing us to compute all quantities $\Pr(r_i | C_k)$ for $1 \leq i \leq M$ and $1 \leq k \leq K$.

The behavior of the model (8.6) depends on the number of word clusters defined in the space $\mathcal{L}$. If there are as many classes as words in the vocabulary ($K = M$), no smoothing is introduced. Conversely, if all the words are in a single class ($K = 1$), the model becomes maximally smooth: the influence of specific semantic events disappears, leaving only a broad (and therefore weak) vocabulary effect to take into account. This may in turn degrade the predictive power of the model.

Generally speaking, as the number of word classes C_k increases, the contribution of $\Pr(r_q | C_k)$ tends to increase, because the clusters become more and more semantically meaningful. By the same token, however, the contribution of $\Pr(C_k | \tilde{c}_{q-1})$ for a given $\tilde{c}_{q-1}$ tends to decrease, because the clusters eventually become too specific and fail to reflect the overall semantic fabric of $\tilde{c}_{q-1}$. Thus, there exists a cluster set size where the degree of smoothing is optimal for the task considered (which has indeed been verified experimentally, cf. [10]).

8.3.2 Document Smoothing

Exploiting document clusters D_ℓ, $1 \leq \ell \leq L$, instead of word clusters leads to a similar expansion:

$$\Pr(r_q | \tilde{c}_{q-1}) = \sum_{\ell=1}^{L} \Pr(r_q | D_\ell) \Pr(D_\ell | \tilde{c}_{q-1}). \qquad (8.7)$$

This time, it is the probability $\Pr(r_q | D_\ell)$ which stems from a unit-composition comparison, and can therefore be obtained with the help of (3.6). As for the probability $\Pr(D_\ell | \tilde{c}_{q-1})$, it depends on the "closeness" of $\tilde{c}_{q-1}$ relative to the centroid of document cluster D_ℓ. Thus, it can be obtained through the empirical multivariate distribution induced by the distance derived from (3.5).

Again, the behavior of the model (8.7) depends on the number of document clusters defined in the space $\mathcal{L}$. As the number of document classes D_ℓ increases, the contribution of $\Pr(r_q|D_\ell)$ tends to increase, to the extent that a more homogeneous topic boosts the effects of any related content words. On the other hand, the contribution of $\Pr(D_\ell|\tilde{c}_{q-1})$ tends to decrease, because the clusters represent more and more specific topics, which increases the chance that the pseudodocument $\tilde{c}_{q-1}$ becomes an outlier. Thus, again there exists a cluster set size where the degree of smoothing is optimal for the task considered (cf. [10]).

8.3.3   Joint Smoothing

Finally, an expression analogous to (8.6) and (8.7) can also be derived to take advantage of both word and document clusters. This leads to a mixture probability specified by

$$\Pr(r_q|\tilde{c}_{q-1}) = \sum_{k=1}^{K}\sum_{\ell=1}^{L} \Pr(r_q|C_k, D_\ell)\Pr(C_k, D_\ell|\tilde{c}_{q-1}), \qquad (8.8)$$

which, for tractability, can be approximated as

$$\Pr(r_q|\tilde{c}_{q-1}) = \sum_{k=1}^{K}\sum_{\ell=1}^{L} \Pr(r_q|C_k)\Pr(C_k|D_\ell)\Pr(D_\ell|\tilde{c}_{q-1}). \qquad (8.9)$$

In this expression, the clusters C_k and D_ℓ are as previously, as are the quantities $\Pr(r_q|C_k)$ and $\Pr(D_\ell|\tilde{c}_{q-1})$. As for the probability $\Pr(C_k|D_\ell)$, it again stems from a unit-composition comparison, and can therefore be obtained accordingly.

Any of expressions (8.6), (8.7), or (8.9) can be used to compute (8.5), resulting in several families of hybrid n-gram+LSM language models. In the case $n = 3$, the use of the integrated trigram+LSM language model resulted in a reduction in the word error rate of up to 16% relative to a standard trigram when evaluated on a subset of the *Wall Street Journal* database [10].

CHAPTER 9

Pronunciation Modeling

In Chapters 6–8, the LSM paradigm was applied in its original incarnation, i.e., with units as words, and compositions as documents. But, as noted at the beginning of this monograph, the fairly generic nature of the LSM properties makes the framework amenable to a variety of tasks beyond this original setup. This Chapter explores a recent foray into one such application, which has to do with modeling the pronunciation of words in the language.

9.1 GRAPHEME-TO-PHONEME CONVERSION

Pronunciation modeling is the process of assigning phonemic/phonetic transcriptions to graphemic word forms [13, 56, 152]. Also called grapheme-to-phoneme conversion (GPC), it is of critical importance to all spoken language applications [36, 96]. For most languages, especially English, GPC is a challenging task, because local correspondences between graphemes and phonemes are difficult to encapsulate within a manageable number of general principles [30, 118, 143]. Hence the interest of a global outlook such as provided by LSM. Before getting to that point, however, let us first review some of the conventional approaches to the problem.

9.1.1 Top–Down Approaches

While GPC can be accomplished via a number of different strategies, traditionally top–down approaches have been the most popular. Largely based on inductive learning techniques, they usually involve decision trees [94] or Bayesian networks [112]. The relevant conversion rules are then effectively embodied in the structure of the associated tree or network.

There are various tradeoffs involved in inductive learning. A decision tree, for example, is first grown by iteratively splitting nodes to minimize some measure of spread (e.g., entropy), and then pruned back to avoid unnecessary complexity and/or overfitting. Since question selection and pruning strategy heavily influence traversal granularity and generalization ability, typical tree designs attempt to strike an acceptable balance between the two. The resulting framework is presented in Fig. 9.1. While such techniques exhibit good performance on "conforming"

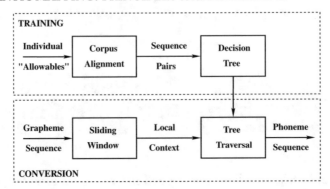

FIGURE 9.1: Top–down grapheme-to-phoneme conversion. (a) Training proceeds using sequence pairs, aligned with the help of language-dependent "allowables," i.e., individual pairs of letters and phonemes which are allowed to correspond [31]. These "allowables" are typically derived by hand, though there is some preliminary evidence that they can also be satisfactorily inferred using a version of the EM algorithm [72, 73].

(b) During actual conversion, a sliding window is used to isolate the local context of each grapheme, and the tree is traversed on the basis of questions asked about this context. When the most likely leaf is reached, the decision tree returns the corresponding phoneme (or phoneme string) associated with the local input grapheme.

words, they tend to degrade rapidly when encountering patterns unusual for the language considered [31].

This often holds true for words of foreign origin, which typically exhibit one or more linguistic phenomena characteristic of another language. Proper names, in particular, are very likely to retain various and sundry idiosyncrasies, due to the (usually) lesser degree of assimilation. As a result, word origin plays an important role in GPC.[1] If only a few exemplars from a given origin are available in the training dictionary, however, it is essentially impossible to extract suitable evidence of alternative regularity. The outcome is a marked degradation in GPC performance.

9.1.2   Illustration

As an illustration, consider the proper name of Indian origin: "*Krishnamoorthy*," for which the phoneme sequence (in SAMPA computer readable phonetic notation [138]):

$$k \quad r \quad I \quad S \quad n \quad @ \quad m \quad U \quad r \quad T \quad i \qquad (9.1)$$

---

[1]In fact, some recent efforts have considered exploiting knowledge of language origin to improve GPC performance in the case of proper names [70].

represents a good estimate of the correct pronunciation. Because this name is not part of the underlying Apple dictionary, the GPC converter previously bundled with MacOS X came up with the (incorrect) sequence

$$k \ r \ I \ S \ n \ \{ \ m \ \mathbf{u} \ 3 \ r \ D \ i \qquad\qquad (9.2)$$

after traversing a dedicated decision tree trained on a relatively large corpus (56K names), albeit predominantly of Western European origin.

Comparing (9.2) with (9.1), three errors stand out: (i) the schwa "@" is replaced by the full vowel "{", (ii) the unvoiced "T" is replaced by the voiced version "D", and (iii) the stressed vowel "U" is replaced by the improper compound "**u** 3". These errors can all be traced to poor generalization properties. Specifically, the ending "3 r D i" results from the influence of a large number of names in the training dictionary ending in "orthy," such as "*Foxworthy.*" The vowel compound comes from the inability of this pattern to account for "oo," hence the awkward attempt to have it both ways by concatenating the two vowels. Finally, the full vowel "{", commonly seen after "n" in names like "*McNamara,*" points to an obvious failure to connect "*Krishnamoorthy*" with the more closely related "*Krishna.*"

9.1.3   Bottom–Up Approaches

Bottom–up systems evolved in an effort to address some of the drawbacks exemplified above. They typically use some form of pronunciation by analogy (PbA) [152]. Given a suitable measure of similarity between words, such as the Levenshtein string-edit distance, they directly retrieve partial pronunciations for local fragments of the input word. Loosely speaking, two words are lexical neighbors if they share a common graphemic substring [152]. The various pronunciation fragments are then concatenated to obtain the final pronunciation. The resulting framework is presented in Fig. 9.2.

I am sending you my resume hope you will consider me for your pretegious company.

This strategy allows for a markedly better handling of rarer contexts, at the expense of a potential loss of robustness [55, 111, 113]. As in top-down systems, there is an implicit need for external linguistic knowledge, in the sense that the methods critically rely on (language-dependent) correspondences between individual graphemes and phonemes. Although recent work has attempted to relax some of these requirements, the handling of "nulls" would seem to remain highly language-dependent [56].

Perhaps even more importantly, this concept offers no principled way of deciding which neighbor(s) of a new word can be deemed to substantially influence its pronunciation. As a case in point, from a Levenshtein distance perspective, the words "*rough,*" "*though,*" and "*through*" would likely be considered lexical neighbors, while in fact they have absolutely no bearing on each other when it comes to the pronunciation of the substring "*ough.*"

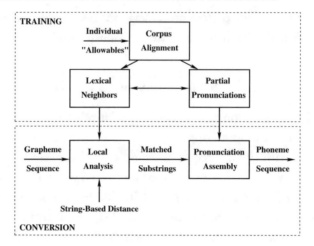

FIGURE 9.2: Bottom–up grapheme-to-phoneme conversion. As before, a local analysis is performed on the input grapheme sequence, but this time a (discrete) string-based measure is used to determine promising lexical neighbors. The idea is to match substrings of the input to substrings of its lexical neighbors, and to convert the various matched substrings to partial pronunciations, again with the help of suitable "allowables." The final pronunciation can then be assembled by concatenating all the partial pronunciations.

9.2   PRONUNCIATION BY LATENT ANALOGY

Pronunciation by latent analogy [13] is an attempt to avoid the PbA pitfalls just mentioned. The need for correspondences between individual graphemes and phonemes is obviated by defining a type of neighborhood which is not specified exclusively in terms of local graphemic substrings, but also, via LSM, relies on global knowledge of the latent relationships between substrings and phonemic information. The resulting framework is presented in Fig. 9.3.

9.2.1   Orthographic Neighborhoods

In this approach, $\mathcal{M}$ is an inventory of letter n-tuples, and $\mathcal{N}$ is a collection of words from a given vocabulary. The LSM framework is used to determine what letter n-tuples are globally most characteristic of words, and map all in-vocabulary words onto the space of all characteristic letter n-tuples. The outcome is a set of *orthographic anchors* (one for each in-vocabulary word), determined automatically from the underlying vocabulary. These orthographic anchors in the LSM space are the equivalent of the semantic anchors of Chapters 6 and 7.

Each out-of-vocabulary word for which a pronunciation is sought is then compared to each orthographic anchor, and the corresponding "closeness" evaluated in the resulting LSM space $\mathcal{L}$. In this case, the closeness measure used is again the composition–composition measure (3.5). If the closeness is high enough, the associated in-vocabulary word is added

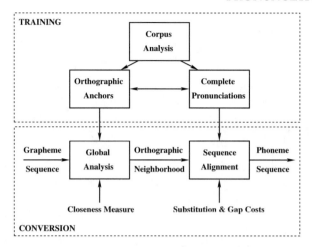

FIGURE 9.3: Pronunciation by latent analogy framework. Pronunciation by latent analogy adopts a generation and selection strategy. LSM is leveraged to generate a set of globally relevant candidates on the grapheme side. Then, among those candidates are picked those pieces that satisfy local constraints on the phoneme side. Because this step uses a data-driven sequence alignment technique [54, 148], the final pronunciation emerges automatically from the alignment itself.

to the so-called *orthographic neighborhood* of the out-of-vocabulary word. For example, in the case of "*Krishnamoorthy*" mentioned above, the orthographic neighborhood does contain the closely related in-vocabulary word "*Krishna*." A simple example of orthographic neighborhood construction is illustrated in Fig. 9.4.

Orthographic neighbors thus obtained can be viewed as the equivalent of similar words in classical PbA [152], except that LSM now defines the concept of similarity in a global rather than a local sense. This means, among other things, that specific matched substrings are no longer available as a byproduct of the lexical match. Once an orthographic neighborhood is available for a given out-of-vocabulary word, however, it is straightforward to gather the corresponding set of pronunciations from the existing dictionary.

9.2.2   Sequence Alignment

By construction, phonetic expansions in this pronunciation neighborhood have the property to contain at least one substring which is "locally close" to the pronunciation sought. Phoneme transcription thus follows via locally optimal sequence alignment and maximum likelihood position scoring. For the name "*Krishnamoorthy*," this process returns the pronunciation

$$ k \quad r \quad I \quad S \quad n \quad @ \quad m \quad u \quad r \quad T \quad i \qquad (9.3) $$

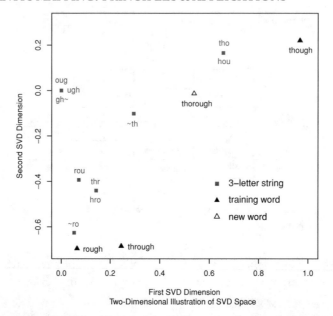

FIGURE 9.4: Orthographic Neighbors in Orthographic Space $\mathcal{L}$ ($R = 2$). Consider the collection of the $N = 3$ words: (i) "*rough*," (ii) "*though*," and (iii) "*through*," for which local graphemic evidence is particularly uninformative. Counting those formed with the word boundary marker ˜, there are $M = 10$ strings of $n = 3$ letters in the unit inventory. Some, for example "*tho*" and "*hou*," always co-occur; "*oug*," "*ugh*," and "*gh˜*" appear in all three words. This leads to the orthographic space $\mathcal{L}$ below. Note the parallels with the semantic inference example of Fig. 7.3, despite the very different choices of $\mathcal{M}$ and $\mathcal{N}$.

While "*oug/ugh/gh˜*" is located at the origin, as expected, letter strings unique to a word score relatively high on the main axes. The orthographic anchors for the three words fall "close" to the letter strings which predict them best. And, not surprisingly, letter strings which are present in two different words indeed appear in the vicinity of both.

Consider now the new word "*thorough*," unobserved in training but consistent with the global graphemic structure. It maps onto the hollow triangle in $\mathcal{L}$. Since this point is closer to (ii) than (iii), the new word is deemed most related to "*though*," despite a closer Levenshtein relationship with "*through*." It thereby inherits the correct pronunciation for the substring "*ough*."

(cf. Table 9.1). A comparison with (9.1) and (9.2) shows that, while still not entirely correct, this expansion solves many of the problems observed in decision tree GPC. In fact, this approach implicitly and automatically takes into account the influence of the entire neighborhood [15], while requiring no external linguistic knowledge.

Compared to classical PbA [56, 152], pronunciation by latent analogy essentially decouples the two underlying tasks of neighborhood generation and pronunciation assembly. Now

TABLE 9.1: Example of Sequence Alignment for *"Krishnamoorthy."*

~	k	r	I	s	~						
~	k	r	I	S	n	@	~				
	~	f	I	S	n	@	n	~			
		~	v	@	m	u	s	~			
			~	m	u	r	z	~			
			~	m	U	r	T	i	~		
	~	m	@	k	{	r	T	i	~		
			~	w	3	r	D	i	~		
~	k	r	I	S	n	@	m	u	r	T	i	~

neighborhood generation involves gathering globally pertinent information on the grapheme side, while pronunciation assembly involves exploiting locally consistent constraints on the phoneme side. This method was observed to be particularly effective on a difficult test collection of proper names with a large diversity of origin [13, 15].

CHAPTER 10

Speaker Verification

In Chapter 9, the LSM paradigm was not applied with the usual LSA-derived choice of $\mathcal{M}$ as a set of words and $\mathcal{N}$ as a collection of documents, but the setup, arguably, was still closely related to the original formulation. This chapter explores a situation where LSM is cast purely as a feature representation mechanism, and the associated features exploited for the purpose of speaker verification. This exemplifies how LSM can also be leveraged for generic pattern recognition.

10.1 THE TASK

Speaker verification is part of the general field of voice authentication. It refers to the process of accepting or rejecting the identity claim of a speaker on the basis of individual information present in the speech waveform [37, 60, 83, 114]. It has received increasing attention over the past two decades, as a convenient, user-friendly way of replacing (or supplementing) standard password-type matching. Many applications involve speaker verification over the telephone; however in what follows we will focus on a desktop voice login task over a far-field microphone [20]. Not only does this entail a particularly challenging acoustic environment, but in addition enrollment data tend to be severely limited. In this type of situation, the authentication methods of interest are those which can concurrently verify both speaker characteristics and verbal content (cf. [83]).

This scenario entails the comparison of the acoustic sequence uttered during recognition (verification utterance) with the aggregated acoustic evidence collected during training (keyphrase-specific reference speaker model). This is typically done using HMM technology with Gaussian mixture distributions (see, e.g., [114, 120]). Conceptually, the verification utterance is aligned against the appropriate subword HMMs constructed from the relevant reference speaker model, and the likelihood of the input speech matching the reference model is calculated. If this likelihood is high enough, the speaker is accepted as claimed. This approach, however, faces scarce data problems for short enrollments. Variance estimation is of particular concern, as the underlying Gaussian mixture distributions run the risk of being too sharp and overfitting the training data [109].

This has sparked interest in an alternative strategy based on the divide and conquer principle. Rather than using a single paradigm to verify both speaker and verbal content simultaneously, it is possible to decouple the two and adopt a different (light-weight) algorithm for each of them: one primarily for global spectral content matching, and the other mostly for local temporal alignment. As it turns out, the LSM paradigm can be exploited to address the former, and simple dynamic time warping (DTW) proves adequate to handle the latter.

10.2   LSM-BASED SPEAKER VERIFICATION

The first step is to extract standard spectral feature vectors (typically every 10ms), using short-term FFT followed by filter bank analysis to ensure a smooth spectral envelope.[1] To represent the spectral dynamics, we also extract, for every frame, the delta and delta–delta parameters. After concatenation, we therefore end up with a sequence of M feature vectors (frames) of dimension N, where, for a typical utterance, $M \approx 200$ and $N \approx 40$.

10.2.1   Single-Utterance Representation

From the above, each utterance is represented by an $M \times N$ matrix of frames, say F, where each row represents the spectral information for a frame and each column represents a particular spectral band over time. Applying the LSM outlook, we can therefore write

$$F = U S V^T, \qquad (10.1)$$

which, for reasons to become clear shortly, we refer to as the decomposition of the utterance into *single-utterance* singular elements U, S, and V.

Such whole utterance representation has been considered before: see, e.g., [2]. The resulting parameterization can be loosely interpreted as conceptually analogous to the Gaussian mixture parameterization in the HMM framework. The main difference is that the Gaussian mixture approach is implicitly based on a subword unit (such as a phoneme), whereas the LSM approach operates on the entire utterance, which introduces more smoothing.

It is intuitively reasonable to postulate that some of the singular elements will reflect more speaker information and some others more verbal content information. But it is not completely clear exactly which reflects what. In [2], a case was made that speaker information is mostly contained in V. Speaker verification was then performed using the Euclidean distance after projection onto the "speaker subspace" defined by V, on the theory that in that subspace utterances from the true speaker have greater measure.

---

[1]A smooth spectral envelope is important to provide a stable representation from one repetition to another of a particular speaker's utterance.

But an equally compelling case could be made under the (dual) assumption that verbal content information is mostly contained in U. In this situation speaker verification could conceivably be performed after projection onto the "content subspace" spanned by U. One would simply compute distances between reference and verification utterances in that subspace, on the theory that a large distance between two utterances with the same verbal content would have to be attributed to a speaker mismatch.

In the standard framework of Chapter 2, such discussion normally leads to a common representation in terms of the row vectors of US and VS: cf. Figs. 2.1(a) and 2.1(b), respectively. The mapping (10.1) defines an LSM space $\mathcal{L}$ which, by definition, captures the major structural associations in F and ignores higher order effects. Closeness in this space is therefore determined by the overall speech patterns observed in the utterance, as opposed to specific acoustic realizations. Specifically, two frames whose representations are close tend to have similar spectral content, and conversely two spectral bands whose representations are close tend to appear in similar frames. The problem, for the present application, is that no firm conclusion can be drawn regarding *where* speaker and content information is concentrated.

Fortunately, given the alternative decoupling strategy mentioned above, this level of detail is not warranted. Because we are not trying to compare individual frames (or spectral bands, for that matter), characterizing individual points in LSM space is unnecessary. In fact, the problem is not so much to analyze a given LSM space as it is to relate distinct LSM spaces to each other. For this, we need to specify a distance measure suitable to compare different global representations. The following justifies and adopts a speaker verification metric specifically tailored to the LSM framework.

10.2.2 LSM-Tailored Metric

Assume, without loss of generality, that (10.1) is associated with a particular training utterance, say the jth utterance, from a given speaker, and consider the set of all training utterances from that speaker. This set will be represented by an $\tilde{M} \times N$ matrix, with $\tilde{M} \approx J M$, where J is the number of training utterances for the speaker. Denoting this $\tilde{M} \times N$ matrix by $\tilde{F}$, it can be decomposed as

$$\tilde{F} = \tilde{U}\,\tilde{S}\,\tilde{V}^T, \tag{10.2}$$

with analogous definitions and properties as in (10.1). In particular, (10.2) defines a similar, though likely distinct, LSM space from the one obtained via (10.1), which was only derived from a single utterance.

Obviously, the set of all training utterances contains the jth utterance, so by selecting the appropriate M rows of $\tilde{F}$, we can define

$$\tilde{F}_{(j)} = F = \tilde{U}_{(j)}\, \tilde{S}\, \tilde{V}^T, \qquad (10.3)$$

where the subscript (j) serves as an index to the jth utterance. Presumably, from the increased amount of training data, the matrices $\tilde{S}$ and $\tilde{V}$ are somewhat more robust versions of S and V, while $\tilde{U}_{(j)}$ relates this more reliable representation (including any embedded speaker information) to the original jth utterance. We refer to (10.3) as the decomposition of the utterance into *multiple-utterance* singular elements $\tilde{U}_{(j)}$, $\tilde{S}$, and $\tilde{V}$, and similarly to the LSM space associated with (10.2) as the underlying *multiple-utterance* LSM space.

This opens up the possibility of relating the two LSM spaces to each other. The equality

$$\tilde{U}_{(j)}\, \tilde{S}\, \tilde{V}^T = U\, S\, V^T, \qquad (10.4)$$

follows from (10.1) and (10.3). To cast this equation into a more useful form, we now make use of the (easily shown) fact that the matrix $(V^T \tilde{V})$ is (both row- and column) orthonormal. After some algebraic manipulations, we eventually arrive at the expression

$$\tilde{S}\left(\tilde{U}_{(j)}^T\, \tilde{U}_{(j)}\right)\tilde{S} = (V^T\tilde{V})^T\, S^2\, (V^T\tilde{V}). \qquad (10.5)$$

Since both sides of (10.5) are symmetric and positive definite, there exists a $(R \times R)$ matrix $D_{j|\tilde{s}}$ such that

$$D_{j|\tilde{s}}^2 = \tilde{S}\left(\tilde{U}_{(j)}^T\, \tilde{U}_{(j)}\right)\tilde{S}. \qquad (10.6)$$

Note that, while $\tilde{U}^T\tilde{U} = I_R$, in general $\tilde{U}_{(j)}^T\tilde{U}_{(j)} \neq I_R$. Thus $D_{j|\tilde{s}}^2$ is closely related, but not equal, to $\tilde{S}^2$. Only as the single-utterance decomposition becomes more and more consistent with the multiple-utterance decomposition does $D_{j|\tilde{s}}^2$ converge to $\tilde{S}^2$.

Taking (10.6) into account and again invoking the orthonormality of $(V^T\tilde{V})$, the equation (10.5) is seen to admit the solution

$$D_{j|\tilde{s}} = (V^T\tilde{V})^T\, S\, (V^T\tilde{V}). \qquad (10.7)$$

Thus, the orthornormal matrix $(V^T\tilde{V})$ can be interpreted as the rotation necessary to map the single-utterance singular value matrix obtained in (10.1) onto (an appropriately transformed version of) the multiple-utterance singular value matrix obtained in (10.2). Clearly, as V tends to $\tilde{V}$ (meaning U also tends to $\tilde{U}_{(j)}$) the two sides of (10.7) become closer and closer to a diagonal matrix, ultimately converging to $S = \tilde{S}$.

The above observation suggests a natural metric to evaluate how well a particular utterance j is consistent with the (multiple-utterance) speaker model: compute the quantity

$D_{j|\tilde{s}} = (V^T \tilde{V})^T S (V^T \tilde{V})$, per (10.7), and measure how much it deviates from a diagonal matrix. For example, one way to measure the deviation from diagonality is to calculate the Frobenius norm of the off-diagonal elements of the matrix $D_{j|\tilde{s}}$.

This in turn suggests an alternative metric to evaluate how well a verification utterance, uttered by a speaker ℓ, is consistent with the (multiple-utterance) model for speaker k. Indexing the single-utterance elements by ℓ, and the multiple-utterance elements by k, we define

$$D_{\ell|k} = (V_\ell^T \tilde{V}_k)^T S_\ell (V_\ell^T \tilde{V}_k), \tag{10.8}$$

and again measure the deviation from diagonality of $D_{\ell|k}$ by calculating the Frobenius norm of its off-diagonal elements. By the same reasoning as before, in this expression the matrix $(V_\ell^T \tilde{V}_k)$ underscores the rotation necessary to map S_ℓ onto (an appropriately transformed version of) $\tilde{S}_k$. When V_ℓ tends to $\tilde{V}_k$, $D_{\ell|k}$ tends to $\tilde{S}_k$, and the Frobenius norm tends to zero. Thus, the deviation from diagonality can be expected to be less when the verification utterance comes from speaker $\ell = k$ then when it comes from a speaker $\ell \neq k$. Clearly, this distance measure is better tailored to the LSM framework than the usual Euclidean (or Gaussian) distance. It can be verified experimentally that it also achieves better performance.

The LSM component thus operates as follows. During enrollment, each speaker $1 \leq k \leq K$ to be registered provides a small number J of training sentences. For each speaker, the enrollment data are processed as in (10.2), to obtain the appropriate right singular matrix $\tilde{V}_k$. During verification, the input utterance is processed as in (10.1), producing the entities S_ℓ and V_ℓ. Then $D_{\ell|k}$ is computed as in (10.8), and the deviation from diagonality is calculated. If this measure falls within a given threshold, then the speaker is accepted as claimed. Otherwise, it is rejected.

10.2.3 Integration with DTW

The LSM approach deliberately discards a substantial amount of available temporal information, since it integrates out frame-level information. Taking into account the linear mapping inherent in the decomposition, it is likely that the singular elements only encapsulate coarse time variations, and smooth out finer behavior. Unfortunately, detecting subtle differences in delivery is often crucial to thwarting noncasual impostors, who might use their knowledge of the true user's speech characteristics to deliberately mimic his or her spectral content. Thus, a more explicit temporal verification should be added to the LSM component to increase the level of security against such determined impersonators.

We adopt a simple DTW approach for this purpose. Although HMM techniques have generally proven superior for time alignment, in the present case the LSM approach already contributes to spectral matching, so the requirements on any supplementary technique are less

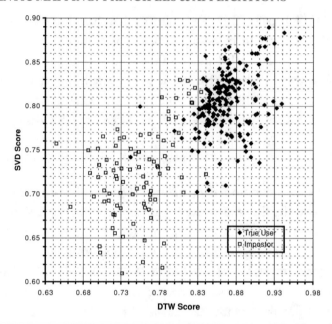

FIGURE 10.1: Performance Space of LSM+DTW Approach. Experiments were conducted using a set of 93 speakers, $K = 48$ true users and $K' = 45$ impostors. True users enrolled by speaking their keyphrase $J = 4$ times. They also provided four instances of a voice login attempt, collected on different days. This resulted in a total of 191 true test utterances, across which the minimum, average, and maximum sentence length were 1.2, 1.8, and 3 sec, respectively.

To increase the severity of the test, each impostor was dedicated to a particular speaker, and was selected on the basis of his/her apparent "closeness" to that user, as reflected in his/her speech characteristics. For example, to impersonate a male speaker who grew up in Australia, we chose another male speaker with an Australian accent. Further, each impostor was given access to the original enrollment keyphrases from the true speaker, and was encouraged to mimic delivery as best as s/he could. This was to reflect the high likelihood of deliberate imposture in desktop voice login, where the true user is typically known to the impostor. Each impostor provided two distinct attempts, for a total of 90 impostor test utterances.

For the appropriate combination of thresholds, the system leads to 0 false acceptances and 20 false rejections (10.4%). After tuning to obtain an equal number of false acceptances and false rejections, we observed approximately a 4% equal error rate.

severe. As it turns out, here DTW suffices, in conjunction with the LSM component, to carry out verbal content verification.

The DTW component implements the classical dynamic time warping algorithm (cf., e.g., [3]). During training, the J training utterances provided by each speaker are "averaged" to define a representative reference utterance s_{avg}. This is done by setting the length of s_{avg} to the average length of all J training utterances, and warping each frame appropriately to come up with the reference frame at that time. During verification, the input utterance, say s_{ver}, is acquired and compared to the reference model s_{avg}. This is done by aligning the time axes of s_{ver} and s_{avg}, and computing the degree of similarity between them, accumulated from the beginning to the end of the utterance on a frame by frame basis. Various distance measures are adequate to perform this step, including the usual Gaussian distance. As before, the speaker is accepted as claimed only if the degree of similarity is high enough.

The integrated system therefore generates two scores for each verification utterance: the deviation from diagonality score from the LSM component, and the degree of similarity from the DTW component. Several options are available to combine the two components. For example, it is possible to combine the two scores into a single one and base the accept/reject decision on that single score. Alternatively, one can reach a separate accept/reject decision for each component and use a voting scheme to form the final decision. Figure 10.1 reports the typical outcome for a difficult task involving deliberate impostors. It shows that using LSM for global spectral matching and DTW for local temporal alignment works well for the (language-independent) task considered.

CHAPTER 11

TTS Unit Selection

The two previous chapters have considered choices of $\mathcal{M}$ and $\mathcal{N}$ increasingly further away from the word-document setup of the original formulation. This chapter explores a recent foray into a even more general application, in the context of concatenative TTS synthesis. This exemplifies how LSM can also be leveraged in generic signal processing problems.

11.1 CONCATENATIVE SYNTHESIS

In modern concatenative TTS synthesis, unit selection identifies which speech segment from a large prerecorded database is the most appropriate at any given point in the speech waveform [5, 14, 29, 38, 51, 87, 144]. As these segments are extracted from disjoint phonetic contexts, discontinuities in spectral shape as well as phase mismatches tend to occur at segment boundaries. Because such artifacts have a deleterious effect on perception, it is important to ensure that such discontinuities are minimized. This in turn requires a high fidelity discontinuity metric for characterizing the acoustic (dis)similarity between two segments [95].

Qualitatively, speech perception is well understood: for example, unnatural sounding speech typically arises from (i) phase mismatches due to interframe incoherence, and (ii) discontinuities in the formant frequencies and in their bandwidths [65]. So at first glance the problem may appear simple and the solution easy to test for success. But in fact, quantitative measures of perceived discontinuity between two segments have proven difficult to agree upon.

This is possibly because user perception of discontinuity may vary, but also because such measures are so intricately tied to the underlying representation of speech. The latter may involve such distinct entities as FFT amplitude spectrum, perceptual spectrum, LPC coefficients, melfrequency cepstral coefficients (MFCC), multiple centroid coefficients, formant frequencies, or line spectral frequencies, to name but a few [95, 146, 151]. While they are all derived from the same Fourier analysis of the signal, each representation has led to its own distance metric to assess spectral-related discontinuities (cf. Fig. 11.1). In contrast, phase mismatches are usually glossed over, to be compensated for belatedly at the signal modification

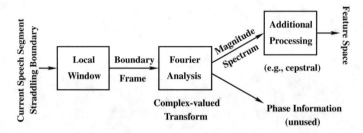

FIGURE 11.1: Conventional feature extraction framework. For each frame on either side of the boundary, a standard Fourier analysis produces the magnitude spectrum of the signal, while phase information is basically discarded. Optional manipulation then yields one of many spectrum-derived feature representations, such as the cepstrum. Finally, the selected representation leads to a specific spectral-related metric, such as Euclidean formant distance, symmetric Kullback–Leibler distance, partial loudness, Euclidean distance between MFCC, likelihood ratio, or mean-squared log-spectral distance, to name but a few.

Many of the above spectral measures have been systematically reviewed in the literature: see, for example, [43, 46, 61, 142, 146, 151]. If there is a single conclusion that can be drawn from these comparisons, it is that no single spectral distance was found to be best in all studies [147]. Not coincidentally, all fall short of ideal performance: none of them succeeds in achieving a correlation with perception greater than 60–70% (cf. [95]).

stage [141]. The approach in [87], for example, involves the cepstral distance at the point of concatenation, and subsequent signal "blending" to address abrupt discontinuities.

11.2   LSM-BASED UNIT SELECTION

This has sparked interest in different signal representations, amenable to alternative discontinuity metrics for characterizing the acoustic (dis)similarity between two segments. In particular, we have recently investigated [14, 17] the feasibility of exploiting the LSM paradigm for that purpose.

11.2.1 Feature Extraction

In that work, LSM is used as a mechanism for feature extraction. Because the resulting transform framework is better suited to preserve globally relevant properties in the region of concatenation, this approach proves beneficial when comparing concatenation candidates against each other [14].

The implementation follows the procedure depicted in Fig. 11.2. For a given boundary point, we define a boundary region in the vicinity of this point which is composed of the K pitch periods before and after the boundary. We then proceed to gather all such pitch periods

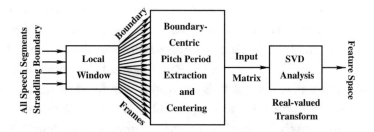

FIGURE 11.2: LSM-based feature extraction framework. LSM-based feature extraction is motivated by two observations. First, implicit or not, the treatment of phase is likely to be facilitated if we consider pitch synchronous epochs. While pitch synchronicity by itself is no panacea in the traditional Fourier framework, largely due to imperfect estimation, it is certainly worth adopting in any effort to expose general patterns in the signal. Besides, it is only at the boundaries that we want to measure the amount of discontinuity, so all the relevant information is likely to be contained within just a few pitch periods surrounding each boundary.

The second observation has to do with the global scope of the analysis. When trying to decide which candidate unit is optimal at any given boundary point, all speech units straddling the boundary are likely to be germane to the decision. Thus, modes should be exposed based on features extracted, not from an individual instance, but from the entire boundary region. Hence the attractiveness of a global optimization framework such as offered by LSM.

for all instances from the database which straddle the boundary. This leads to a matrix W where each row corresponds to a particular (centered[1]) pitch period near the given boundary. LSM then performs a modal (eigen)analysis via a pitch synchronous SVD in each boundary region of interest. In this application, $\mathcal{M}$ is therefore an inventory of centered pitch periods from the boundary region, and $\mathcal{N}$ is a collection of time slices, each comprising the relevant time samples from the pitch period inventory.

11.2.2 Comparison to Fourier Analysis

The resulting decomposition acts as a data-driven alternative to traditional Fourier analysis. To see that, denote the sinusoidal transform kernel by the symmetric complex matrix Φ such that $\Phi_{k\ell} = (1/\sqrt{N}) \exp\{-j2\pi k\ell/N\}$. In classical Fourier analysis, this kernel is applied to the ith row r_i of the matrix W to yield

$$X_i = r_i \, \Phi, \tag{11.1}$$

---

[1]With a *centered* representation, the boundary can be precisely characterized by a single vector in the resulting feature space [17]. (In a more conventional framework, the boundary is normally inferred a posteriori from the position of the two vectors on either side.)

where X_i is the (normalized) Fourier transform vector associated with r_i. In other words, the Fourier-derived features (X_i) are obtained via projection onto a set of signal-independent complex sinusoids (see, e.g., [42]). As the inverse transform yields $r_i = X_i \Phi^H$, where the superscript H denotes Hermitian transposition, Φ is (column)orthornormal just like U and V.

The sinusoidal transform kernel Φ is reasonably well justified from a psycho-acoustic point of view, since the human ear acts as a kind of Fourier spectrum analyzer. On the other hand, the ear most likely is a nonlinear system, whose true "analysis" parameters are yet unknown [137]. In this respect, (11.1) can be regarded as an approximate (linear) analysis of the acoustic signal.

It thus becomes clear that (2.3) simply corresponds to an alternative linear approximation, brought about by another choice of transform kernel. Indeed, from (2.3), each row r_i of W can be expressed as $r_i = u_i S V^T = \bar{u}_i V^T$, which leads to the expression

$$\bar{u}_i = r_i V. \qquad (11.2)$$

Clearly, the inner product of r_i with the kth right singular vector can be interpreted as a measure of the strength of the signal at the mode represented by this right singular vector. Thus, $\bar{u}_i$ can be viewed as the transform vector associated with r_i, given the (data-driven) transform kernel represented by V. Said another way, the new features ($\bar{u}_i$) are derived in terms of a separately optimized set of basis components for each boundary region of interest.

11.2.3 Properties

We readily acknowledge that the SVD kernel (V) is most likely inferior to the Fourier kernel as a general-purpose signal analysis tool, if only because it does not explicitly expose the concept of frequency. However, its use has several benefits:

- the singular vectors are, by construction, inherently tailored to the boundary region considered, in contrast with the traditional set of signal-independent complex sinusoids;

- the kernel offers a global view of what is happening in this boundary region, as encapsulated in the vector space $\mathcal{L}$;

- this representation is parsimonious, to the extent that an empirically consistent value is selected for the dimension R of the space; and

- since this is a real-valued transform, both amplitude and phase information are retained, and in fact contribute simultaneously to the outcome.

In the present application, these properties make the LSM transform domain an attractive alternative to traditional Fourier processing.

In summary, the LSM approach leads to an efficient, optimized (in the least-squares sense), boundary-centric representation of the problem. Given the unit–unit comparison metric (3.2), this in turn yields an alternative measure of discontinuity across the boundary, of the form

$$d(S_1, S_2) = \tilde{K}(\text{original } S_1) + \tilde{K}(\text{original } S_2) - \tilde{2}K(\text{concatenated } S_1\text{-}S_2). \qquad (11.3)$$

where S_1 and S_2 represent two speech segments from the database, and $\tilde{K}(\cdot)$ is a function of the closeness $K(\cdot)$ of (3.2) between individual vectors across the boundary. In fact, several discontinuity metrics in $\mathcal{L}$ can be derived to express cumulative differences in closeness before and after concatenation (cf. [17]). An experimental evaluation of these metrics was conducted in [17] using a voice database deployed in MacinTalk, Apple's TTS offering on MacOS X [23]. This comparison underscored a better correlation with perceived discontinuity, as compared with a widely used baseline measure [41, 50]. This confirms the viability and effectiveness of the LSM framework for TTS unit selection.

11.3   LSM-BASED BOUNDARY TRAINING

Boundary-centric feature extraction can also be leveraged for unit segmentation when composing the underlying unit inventory. The task is to systematically optimize all unit boundaries *before* unit selection, so as to effectively minimize the likelihood of a really bad concatenation. We refer to this (offline) optimization as the data-driven "training" of the unit inventory, in contrast to the (run time) "decoding" process embedded in unit selection [16]. This guarantees that at run time, uniformly high quality units are available to choose from.

In such training, the evaluation criterion (11.3) is embedded in an iterative procedure to sequentially refine the unit boundaries. The basic idea is to focus on each possible boundary region in turn, compute the LSM space associated with this region, adjust individual boundaries in that space, update the boundary region accordingly, and iterate until convergence. At each iteration, the discontinuity score (11.3) resulting from the concatenation of every instance of a particular unit with all other instances of that unit is computed for a neighborhood of the current hypothesized boundary. The cut point yielding the lowest average score is then retained as the new boundary for the next iteration.

The iterative boundary training procedure follows the flowchart of Fig. 11.3. The initialization step can be performed in a number of different ways; but in practice, we have found little difference in behavior based on these various forms of initial conditions [16]. For example, the initial boundary for each instance can be placed in the most stable part of the phone (where the speech waveform varies the least), or, more expediently, simply at its midpoint [16]. Since the boundary region shifts from one iteration to the next, the LSM space does not stay static. While this complicates the derivation of a theoretical proof of convergence, it can still be done by exploiting the fact that after each iteration the space remains relatively close to its previous

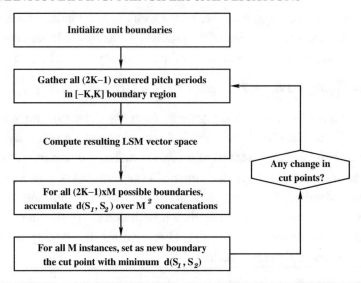

FIGURE 11.3: Iterative Training of Unit Boundaries. Starting from suitable initial conditions, we gather the $2K - 1$ centered pitch periods for each unit instance, and derive the resulting LSM space $\mathcal{L}$. This leads to $(2K - 1)M$ feature vectors in the space, and hence as many potential new boundaries. For each of them, we compute the associated average discontinuity by accumulating (11.3) over the set of M^2 possible concatenations. This results in $2K - 1$ discontinuity scores for each instance, the minimum value of which yields the cut point to be retained. The new boundaries form the basis for a new boundary region, and the procedure iterates until no change in cut points is necessary.

incarnation. As shown in [18], the iterative procedure does converge in the least squares sense to a global minimum.

The associated final boundaries are therefore globally optimal across the entire set of observations for the phoneme considered. Note that, with the choice of the LSM framework, this outcome holds given the exact same discontinuity measure later used in unit selection. Not only does this result in a better usage of the available training data, but it also ensures tightly matched conditions between training and decoding. Systematic listening tests confirmed that utterances synthesized using this approach tend to comprise less egregious discontinuities than those synthesized in a more conventional way. This again underscores the viability of LSM-based feature extraction in accounting for the perception of discontinuity between acoustic units.

Part III

Perspectives

CHAPTER 12

Discussion

As evidenced by Chapters 9–11, the LSM paradigm has wider applicability than initially contemplated in the original word-document incarnation of Chapters 6–8. Generalizing the formulation to generic units and compositions enables the underlying mapping to potentially capture hidden patterns in all manners of large collections. This opens tantalizing possibilities in many areas, both within and beyond natural language processing.

12.1 INHERENT TRADEOFFs

How effective LSM will be in these various domains, however, hinges on the various tradeoffs associated with the approach. These tradeoffs fall into three main categories: descriptive power, domain sensitivity, and adaptation capabilities.

12.1.1 Descriptive Power

The descriptive power of LSM is subordinated to the particular constraints intrinsic to LSM modeling. One such constraint is the L_2 norm arising from (2.4). Since there is no reason to believe it is the best choice in all situations, the generic optimality of the LSM framework can be debated.

Consider the case of linguistic phenomena. Depending on many subtly intertwined factors such as frequency and recency, linguistic co-occurrences may not always have the same interpretation. When comparing word counts, for example, observing 100 occurrences versus 99 is markedly different from observing one versus zero. Relying on squared deviation may therefore mask potentially important aspects of the problem. In information retrieval, this has motivated the investigation of an alternative objective function based on the Kullback–Leibler divergence [84], which in turn led to the PLSA framework reviewed in Chapter 5.

This approach has the advantage of providing an elegant probabilistic interpretation of (2.3), at the expense of requiring a conditional independence assumption on the words and the documents, given a hidden variable representing the topic of discourse [85]. As we discussed in

Chapter 5, unfortunately, this independence assumption is almost always violated in practice.[1] For example, there is no reason to believe that, given the topic of financial news, words such as "stock" or "bond" would be statistically independent of the particular documents in which they appear: after all, some articles routinely focus on the stock market, and are therefore less likely to mention "bond," while others focus on the bond market, with the opposite outcome.

Another fundamental constraint stems from the elementary "bag-of-units" modeling, which is by nature unable to capitalize on the local constraints present in the composition language. This is fine in applications, like information retrieval, where local constraints are discarded anyway. But in some situations, forgoing local constraints are simply not acceptable: in spoken interface control, for example, we have seen that different commands could map onto the same point in LSM space if word order were completely disregarded (cf. Chapter 7). This forces somewhat awkward extensions such as word agglomeration, a solution not necessarily feasible in all environments [22].

Such optimality and modeling limitations can be viewed as instances of the familiar tradeoff between descriptive power and mathematical tractability.

12.1.2 Domain Sensitivity

Like all data-driven techniques, LSM is inherently dependent on the quality of the training data, expressed here in terms of the particular unit-composition co-occurrences observed in the collection $\mathcal{N}$. Basically, pertinent patterns not present in $\mathcal{N}$ cannot be captured. This entails a relatively high sensitivity to the general domain sampled during training, and accordingly restricts cross-domain applicability.

In language modeling, for instance, the framework exhibits a fairly narrow semantic specificity, in the sense that the space $\mathcal{L}$ does not appear to reflect any of the pragmatic characteristics of the task considered. Hence a relatively high sensitivity to the style of composition [10]. Perhaps the LSM framework would benefit from explicitly including a "composition style" component. In [119], for example, it has been suggested to define an $(M \times M)$ stochastic matrix (a matrix with nonnegative entries and row sums equal to 1) to account for the way style modifies the frequency of units. This solution, however, makes the assumption—not always valid—that this influence is independent of the underlying subject matter.

Further contributing to domain sensitivity is *polysemy*, the fact that some units may have multiple interpretations in the composition language. This phenomenon was already mentioned in conjunction with the clustering instances presented in Fig. 1.2. More generally,

---

[1]Incidentally, as we briefly mentioned in Chapter 5, more sophisticated models have since been proposed (cf., e.g., [33]). They tend, however, to rely on somewhat arcane text generation scenarios that are similarly difficult to justify.

the problem is that the same unit may occasionally co-occur with several different, and unrelated, sets of compositions. In that situation, the associated representation of the unit in the LSM space typically falls "between clusters," i.e., (i) its position in $\mathcal{L}$ tends to be unsatisfactory, and (ii) it is liable to contaminate other clusters in the vicinity. When practical to do so, it is therefore recommended to systematically label multiple meanings beforehand, to keep instances of polysemy to a minimum.

12.1.3 Adaptation Capabilities

The last inherent tradeoff is tied to the online adaptation of the LSM framework. At issue is LSM recomputation as new data become available. Clearly, it would be desirable to update the LSM space to reflect any new information uncovered. The scale of the data in a specific application, however, obviously impacts the amount of computations involved (cf. Chapter 4). At the present time, recomputing a very large SVD on the fly is usually not practical. In such cases, there is an implicit assumption that actual usage will only involve new compositions which closely conform to the training collection. It thus becomes all the more critical to have a sufficient amount of training data upfront in order to achieve an acceptable level of performance.

This limitation has been extensively analyzed in the context of semantic inference [22], where it was found that, in particular, complete outcome coverage was required to avoid systematic errors. In practice, this requires recomputing the LSM space each time a new outcome is added. On that subject, we note that some adaptive techniques have been proposed, including a linear transform solution based on Cholesky factorizations [12]. Unfortunately, they appear to be limited in their applicability, and in particular cannot compete with recomputation when the amount of new data is large.

Of course, it is also possible to rely on some of the incremental SVD implementations discussed in Chapter 4, such as out-of-core techniques [110] and parallel and orthogonal decompositions [34]. As already mentioned, however, these approaches still require careful monitoring of the incremental loss of orthogonality, and regular recomputation to compensate for it.

12.2 GENERAL APPLICABILITY

By staying cognizant of the above LSM limitations, it is likely possible to leverage the framework in a wide spectrum of applications, not just in natural language processing, but also in additional areas of signal processing and pattern recognition.

12.2.1 Natural Language Processing

Chapter 1 has reviewed how the LSM framework (in its LSA incarnation) can be applied to information retrieval, as well as word and document clustering. Chapters 6–8 have discussed in detail how LSM can be leveraged for more general semantic classification and statistical language modeling, and reviewed successful applications such as junk e-mail filtering, desktop interface control, and hybrid semantic language modeling for speech recognition. All of these tasks exploit the dual mapping of units and compositions into the LSM space, which in this case exposes the semantic relationships between words and documents.

Clearly, many more applications in natural language processing can potentially benefit from this semantic representation, however simplified it may be. Particularly promising are the related areas of story segmentation, title generation, and document summarization. In story segmentation, the task is to break up the document into meaningful building blocks, such as a set of coherent paragraphs [82]. Title generation then assigns titles to these building blocks, or more generally extracts key concepts and relationships present in the document [150]. And as these concepts are ordinarily woven into topics in the discourse, document summarization produces a concise abstract of suitable length [133].

Various efforts are currently under way to explore how the LSM framework can be applied to these areas. For example, in story segmentation, LSM has been considered to complement other models [32]. Recent title generation approaches have investigated how the association between titles and documents could be rendered in LSM space [107]. And in document summarization, current directions are centered around the integration of independent clues, potentially including LSM features [71].

12.2.2 Generic Pattern Recognition

To conclude this chapter, we offer a case study of how LSM might be applied to a more general pattern recognition problem: the identification of the character encoding a computer file might use, also referred to as encoding detection. Encoding information is critical to resolving such questions as which application(s) can open the file, what fonts are required for optimal display, etc. Typically, it is specified as part of special attributes, collectively functioning as an internal label, included in the file header. However, this solution assumes a very high level of cooperation between file creation and later processing, with often negative consequences on interoperability. In addition, it suffers from a serious lack of robustness in the face of even mild corruption. Hence the attractiveness of detecting character encodings based on the content of the file itself.

Adopting an LSM outlook, we can postulate that any encoding is characterized not just by the presence or absence of certain bytes, but by the *co-occurrences* of certain bytes with others. Thus encoding detection becomes a matter of byte co-occurrence analysis, and the LSM framework can be leveraged in a manner similar to semantic classification. Assuming we collect a suitable corpus of variously encoded files, each encoding can be represented by an *encoding anchor* in the LSM space, and each incoming file can then be classified against these anchor points. If deemed close enough to a particular anchor point, the file is assigned the associated

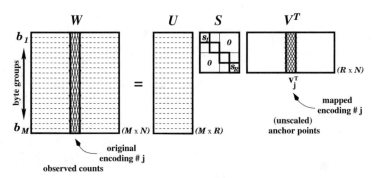

FIGURE 12.1: Finding the encoding anchors. While the input data are much different, the framework remains exactly the same as in Fig. 2.2. In this application, M is on the order of 10,000, N approximately 50, and n chosen to be 2 or 3. Though no "semantic" interpretation of the LSM mapping applies in this case, it remains that the relative positions of the anchors are determined by the overall patterns observed in $\mathcal{T}$, as opposed to specific constructs. This increases the robustness of the encoding detector.

encoding. Otherwise, the file is tagged as unknown, or possibly corrupted, and passed to the user for disambiguation.

In such a scenario, $\mathcal{T}$ is the database of N training files, and $\mathcal{V}$ the set of all M groups of $1, 2, \ldots, n$ bytes observed in $\mathcal{T}$. During training, we decompose the matrix of co-occurrences as shown in Fig. 12.1, the byte-file version of Fig. 2.2. The mapping then is between each original file exemplifying a particular encoding, and the R-dimensional vector $\bar{v}_j = v_j S$ representing this encoding in the LSM space $\mathcal{L}$.

To evaluate the feasibility of such an approach, preliminary experiments were conducted using the set of (UTF-16BE encoded) `*.strings` files contained in the `*.lproj` folders of the directory `/System/Framework` under MacOS X. All `*.strings` files gathered for each language were first converted to the various multiple encodings each language supports.[2] Files associated with UTF-8 and UTF-16 encodings were then separated into different documents based on the underlying script, e.g. UTF-8_ROMAN, etc. Finally, each file was tokenized into a series of bytes, and the matrix W created using the value $n = 2$. In another words, script-encoding pair information was used to define the anchor points. Overall, the results obtained confirmed the viability of the method as an automatic detector of arbitrary character encodings.

---

[2]For example, all `French.lproj/*.strings` files were duplicated as necessary to create UTF-8 and X-MAC-ROMAN versions. Similarly, all `Japanese.lproj/*.strings` files were duplicated to create UTF-8, X-MAC-JAPANESE, EUC-JP and ISO2022-JP encodings.

CHAPTER 13

Conclusion

To conclude the monograph, this Chapter provides a summary of the material covered, as well as some final perspectives on LSM.

13.1 SUMMARY

This monograph has described latent semantic mapping, a data-driven framework for modeling globally meaningful relationships implicit in large volumes of data. This framework is a generalization of a paradigm originally developed to capture hidden word patterns in a text document corpus. Over the past decade, this paradigm has proven effective in an increasing variety of fields, gradually spreading from query-based information retrieval to word clustering, document/topic clustering, automated call routing, and several other speech and language processing applications.

This success can be largely attributed to three unique characteristics:

- the mapping of discrete entities (units and compositions) onto a continuous parameter space, where efficient machine learning algorithms can be applied;

- the dimensionality reduction inherent in the process, which makes complex natural language problems tractable; and

- the intrinsically global outlook of the approach, which tends to complement the local optimization performed by more conventional techniques.

By taking advantage of these properties, it is possible to uncover useful structural patterns in the underlying collection of compositions, patterns which are otherwise partially obscured by the randomness of unit choice with respect to information extraction.

In many natural language applications, such as information retrieval, units are words, and compositions are documents. With this setup, the LSM paradigm leads to a convenient represention of latent semantic relationships between these various entities. This in turn can be leveraged in multiple applications, which we have exemplified in the domains of junk e-mail filtering, semantic inference for spoken interface control, and large vocabulary speech recognition language modeling.

But the fairly generic nature of the LSM properties also makes the framework applicable to other areas, which we have exemplified in the domains of pronunciation modeling, speaker verification, and concatenative TTS synthesis. We have also discussed potential applicability to even more general pattern recognition and signal processing problems.

13.2   PERPECTIVES

The flexibility of the LSM framework, along with its three main properties of continuous embedding for discrete entities, built-in dimensionality reduction, and globally meaningful topological structure, makes it attractive for data-driven information extraction in a potentially wide range of problems. Of course, one has to stay cognizant of the bounds of the framework. How effective LSM is to a particular application hinges on four caveats:

- the narrow semantic specificity of the co-occurrence approach, which is vulnerable to such phenomena as composition style and polysemy;

- the "bag-of-units" modeling characteristic of the framework, which, by not exposing local context, may sometimes be a drawback;

- the L_2 optimality criterion underlying the SVD, which may not be appropriate in certain situations; and

- the computational complexity of updating the LSM space as new data becomes available, despite on-going efforts to derive efficient algorithms for this purpose.

Because these limitations are inherent to the LSM paradigm, they collectively circumscribe its range of applicability.

In practical terms, each of the above caveats has a deleterious impact on the quality of the LSM space representation. Polysemy typically dilutes cluster homogeneity; any domain and/or style mismatch between training and testing tends to degrade performance; in some applications, the inability to model local relationships forces "function" units[1] to be handled separately;[2] and incremental SVD (or alternative decomposition) techniques require thorough validation, lest any loss of orthogonality entail hard-to-detect numerical errors.

Fortunately, these inherent LSM limitations can be somewhat mitigated through careful attention to the expected task, domain of use, and size of the training collection. For example, instances of polysemy can be kept to a minimum by systematically labelling multiple meanings

---

[1]This terminology refers to units that tend to be more meaningful at the local than the global level. In natural language applications, for example, function words normally have more of a syntactic (local) than semantic (global) role in the given document.

[2]This is the case, for example, in language modeling for speech recognition, where substitutions, insertions, and deletions associated with function words typically account for a large proportion of the errors.

beforehand. Also, it is often effective to incorporate LSM into a more comprehensive paradigm which integrates higher-level structure and knowledge. The integrated solution then comprises components relying on diverse, and hopefully complementary, sources of information, as is done in multi-span language modeling.

In that sense, the advantages of LSM substantially outweigh its drawbacks. Clearly, more work is necessary before claiming to satisfactorily encapsulate all the different facets of latent structure in any composition language (let alone, of course, a natural language like English). Yet, LSM is likely to have wide applicability in any situations involving discrete entities, complex parameterizations with high dimensionality, and conventional algorithms focused on local phenomena. It is hoped that the discussion of the various benefits and trade-offs of LSM presented in this monograph will stimulate application to the potentially numerous areas (both within and beyond natural language processing) involving such situations.

Bibliography

[1] I. Androutsopoulos, G. Paliouras, V. Karkaletsis, G. Sakkis, C. D. Spyropoulos and P. Stamatopoulos, "Learning to filter spam e-mail: a comparison of a naive Bayesian and a memory-based approach," in *Proc. Workshop Mach. Learning Textual Info. Access*, Lyon, France, pp. 1–13, 2000.

[2] Y. Ariki and K. Doi, "Speaker recognition based on subspace methods," in *Proc. Int. Conf. Spoken Language Proc.*, Yokohama, Japan, pp. 1859–1862, Sept. 1994.

[3] K. T Assaleh, K. R. Farrell, M. S. Zilovic, M. Sharma, D. K. Naik and R. J. Mammone, "Text–dependent speaker verification using data fusion and channel detection," in *Proc. SPIE*, Vol. 2277, San Diego, CA, pp. 72–82, Aug. 1994.

[4] L. R. Bahl, F. Jelinek, and R. L. Mercer, "A Maximum Likelihood Approach to Continuous Speech Recognition," *IEEE Trans. Pattern Anal. Mach. Intel.*, Vol. PAMI–5, No. 2, pp. 179–190, March 1983.

[5] M. Balestri, A. Pachiotti, S. Quazza, P. L. Salza, and S. Sandri, "Choose the best to modify the least: a new generation concatenative synthesis system," in *Proc. 6th Eur. Conf. Speech Commun. Technol.*, Budapest, Hungary, pp. 2291–2294, Sept. 1999.

[6] L. E. Baum, "An inequality and associated maximization technique in statistical estimation of probabilistic functions of a Markov process," *Inequalities*, Vol. 3, pp. 1–8, 1972.

[7] J. R. Bellegarda, "Context- dependent vector clustering for speech recognition," in *Speech and Speaker Recognition: Advanced Topics*, C.-H. Lee, F. K. Soong, and K. K. Paliwal, Eds., New York: Kluwer, chapter 6 pp. 133–157, March 1996.

[8] J. R. Bellegarda, "A multi-span language modeling framework for large vocabulary speech recognition," *IEEE Trans. Speech Audio Proc.*, Vol. 6, No. 5, pp. 456–467, Sept. 1998. 10.1109/89.709671

[9] J. R. Bellegarda, "Large vocabulary speech recognition with multi-span statistical language models," *IEEE Trans. Speech Audio Proc.*, Vol. 8, No. 1, pp. 76–84, Jan. 2000. 10.1109/89.817455

[10] J. R. Bellegarda, "Exploiting Latent Semantic Information in Statistical Language Modeling," *Proc. IEEE*, *Spec. Issue Speech Recog. Understanding*, B. H. Juang and S. Furui, Eds., Vol. 88, No. 8, pp. 1279–1296, August 2000.

[11] J. R. Bellegarda, "Robustness in statistical language modeling: review and perspectives," in *Robustness in Language and Speech Technology*, G. J. M. van Noord and J. C. Junqua, Eds. Dortrecht: Kluwer, pp. 113–135, Feb. 2001.

[12] J. R. Bellegarda, "Fast update of latent semantic spaces using a linear transform frame-work," in *Proc. Int. Conf. Acoustics, Speech, Signal Proc.*, Orlando, FL, pp. 769–772, May 2002.

[13] J. R. Bellegarda, "A latent analogy framework for grapheme-to-phoneme conversion," in *Proc. 8th Euro. Conf. Speech Comm. Technol.*, Geneva, Switzerland, pp. 2029–2032, Sept. 2003.

[14] J. R. Bellegarda, "A novel discontinuity metric for unit selection text–to–speech syn-thesis," in *Proc. 5th ISCA Speech Synthesis Workshop*, Pittsburgh, PA, pp. 133–138, June 2004.

[15] J. R. Bellegarda, "Unsupervised, language-independent grapheme–to–phoneme conver-sion by latent analogy," in *Speech Communication*, Amsterdam: Elsevier Science, Vol. 46, No. 2, pp. 140–152, June 2005. 10.1016/j.specom.2005.03.002

[16] J. R. Bellegarda, "LSM–based boundary training for concatenative speech synthesis," in *Proc. Int. Conf. Acoust., Speech, Signal Proc.*, Toulouse, France, May 2006.

[17] J. R. Bellegarda, "A global, boundary-centric framework for unit selection text-to-speech synthesis," *IEEE Trans. Speech Audio Proc.*, Vol. SAP–14, No. 4, July 2006.

[18] J. R. Bellegarda, "Further developments in LSM–based boundary training for unit selection TTS," in *Proc. Int. Conf. Spoken Language Proc.*, Pittsburgh, PA, Sept. 2006.

[19] J. R. Bellegarda, J. W. Butzberger, Y. L. Chow, N. B. Coccaro and D. Naik, "A novel word clustering algorithm based on latent semantic analysis," in *Proc. Int. Conf. Acoust., Speech, Signal Proc.*, Atlanta, GA, pp. I172–I175, May 1996.

[20] J. R. Bellegarda, D. Naik, M. Neeracher and K. E. A. Silverman, "Language–independent, Short-Enrollment voice verification over a far-field microphone," in *Proc. 2001 IEEE Int. Conf. Acoust., Speech, Signal Proc.*, Salt Lake City, Utah, May 2001.

[21] J. R. Bellegarda, D. Naik and K. E. A. Silverman, "Automatic junk e–mail filtering based on latent content," in *Proc. 2003 IEEE Aut. Speech Recog. Understanding Workshop*, St. Thomas, U. S. Virgin Islands, pp. 465–470, Dec. 2003.

[22] J. R. Bellegarda and K. E. A. Silverman, "Natural language spoken interface control using data-driven semantic inference," *IEEE Transactions on Speech and Audio Processing*, Vol. SAP–11, No. 3, pp. 267–277, May 2003. 10.1109/TSA.2003.811534

[23] J. R. Bellegarda, K. E. A. Silverman, K. A. Lenzo and V. Anderson, "Statistical prosodic modeling: from corpus design to parameter estimation," *IEEE Trans. Speech Audio Proc., Special Issue Speech Synthesis*, N. Campbell, M. Macon, and J. Schroeter, Eds. Vol. SAP–9, No. 1, pp. 52–66, Jan. 2001.

[24] M. Belkin and P. Niyogi, "Laplacian eigenmaps and spectral techniques for embed-ding and clustering," in *Advances in Neural Information Processing Systems (NIPS) 14,*

T. Dietterich, S. Becker and Z. Ghahramani, Eds. Cambridge, MA: MIT Press, Sept. 2002.

[25] M. W. Berry, "Large-scale sparse singular value computations," *Int. J. Supercomput. Appl.*, Vol. 6, No. 1, pp. 13–49, 1992.

[26] M. W. Berry, S. T. Dumais and G. W. O'Brien, "Using linear algebra for intelligent information retrieval," *SIAM Rev.*, Vol. 37, No. 4, pp. 573–595, 1995. 10.1137/1037127

[27] M. W Berry and R. D. Fierro, "Low-rank orthogonal decompositions for information retrieval applications," *Numer. Linear Algebr. Appl.*, Vol. 1, No. 1, pp. 1–27, 1996.

[28] M. Berry and A. Sameh, "An overview of parallel algorithms for the singular value and dense symmetric eigenvalue problems," *J. Comput. Appl. Math.*, Vol. 27, pp. 191–213, 1989. 10.1016/0377-0427(89)90366-X

[29] M. Beutnagel, A. Conkie, J. Schroeter, Y. Stylianou and A. Syrdal, "The AT&T next-gen TTS system," in *Proc. 137th Meeting Acoust. Soc. Am.*, pp. 18–24, 1999.

[30] A. W. Black, K. Lenzo and V. Pagel, "Issues in building general letter-to-sound rules," in *Proc. 3rd Int. Workshop Speech Synthesis*, Jenolan Caves, Australia, pp. 77–80, Dec. 1998.

[31] A. W. Black and P. Taylor, "Automatically clustering similar units for unit selection in speech synthesis," in *Proc. 5th Eur. Conf. Speech Commun. Technol.*, Rhodes, Greece, pp. 601–604, Sept. 1997.

[32] D. M. Blei and P. J. Moreno, "Topic segmentation with an aspect hidden markov model," in *Proc. ACM SIGIR Int. Conf. R and D in Information Retrieval*, New Orleans, LA, pp. 343–348, 2001.

[33] D. M. Blei, A. Y. Ng, and M. I. Jordan, "Latent dirichlet allocation," in *Advances Neural Info. Proc. Systems (NIPS) 14*, T. Dietterich, S. Becker, and Z. Ghahramani, Eds., MIT Press, Cambridge, MA, pp. 601–608, September 2002.

[34] M. Brand, "Incremental singular value decomposition of uncertain data with missing values," *MERL Technical Report* TR-2002-24, Cambridge, MA: Mitsubishi Electric Research Laboratory, May 2002.

[35] T. Brants, F. Chen, and I. Tsochantaridis, "Topic-based document segmentation with probabilistic latent semantic analysis," in *Conf. Information and Knowledge Management*, MacLean, VA, pp. 211–218, Nov. 2002.

[36] W. Byrne, M. Finke, S. Khudanpur, J. McDonough, H. Nock, M. Riley, M. Saraclar, C. Wooters and G. Zavaliagkos, "Pronunciation modeling using a hand–labelled corpus for conversational speech recognition," in 1998 *Proc. Int. Conf. Acoust., Speech, Signal Proc.*, Seattle, WA, pp. 313–316, May 1998.

[37] J. P. Campbell Jr., "Speaker recognition: a tutorial," *Proc. IEEE*, Vol. 85, No. 9, pp. 1437–1462, Sept. 1997.

[38] W. N. Campbell and A. Black, "Prosody and the selection of source units for concatenative synthesis," in *Progress in Speech Synthesis*, J. van Santen, R. Sproat, J. Hirschberg and J. Olive, Eds., New York: Springer, pp. 279–292, 1997.

[39] B. Carpenter and J. Chu-Carroll, "Natural language call routing: a robust, self-organized approach," in *Proc. Int. Conf. Spoken Language Proc.*, Sydney, Australia, pp. 2059–2062, Dec. 1998.

[40] X. Carreras and L. Márquez, "Boosting trees for anti-spam email filtering," in *Proc. Int. Conf. Recent Advances Natl. Language Proc.*, Tzigov Chark, Bulgaria, 2001.

[41] P. Carvalho, L. Oliveira, I. Trancoso and M. Viana, "Concatenative speech synthesis for european portuguese," in *Proc. 3rd ESCA Speech Synthesis Workshop*, Jenolan Caves, Australia, pp. 159–163, Nov. 1998.

[42] D. C. Champeney, *A Handbook of Fourier Theorems*, Cambridge: Cambridge University Press, 1987.

[43] D. Chappell and J. H. L. Hansen, "A comparison of spectral smoothing methods for segment concatenation based speech synthesis," *Speech Commun.*, Vol. 36, No. 3–4, pp. 343–373, 2002.

[44] C. Chelba, D. Engle, F. Jelinek, V. Jimenez, S. Khudanpur, L. Mangu, H. Printz, E. S. Ristad, R. Rosenfeld, A. Stolcke and D. Wu, "Structure and performance of a dependency language model," in *Proc. 5th Eur. Conf. Speech Commun. Technol.*, Rhodes, Greece, Vol. 5, pp. 2775–2778, Sept. 1997.

[45] C. Chelba and F. Jelinek, "recognition performance of a structured language model," in *Proc. Sixth Euro. Conf. Speech Comm. Technol.*, Budapest, Hungary, Vol. 4, pp. 1567–1570, Sept. 1999.

[46] J. -D. Chen and N. Campbell, "Objective distance measures for assessing concatenative speech synthesis," in *Proc. 6th Eur. Conf. Speech Commun. Technol.*, Budapest, Hungary, pp. 611–614, Sept. 1999.

[47] J. Chu–Carroll and B. Carpenter, "Dialog management in vector-based call routing," in *Proc. Conf. Assoc. Comput. Linguistics ACL/COLING*, Montreal, Canada, pp. 256–262, 1998.

[48] N. Coccaro and D. Jurafsky, "Towards better integration of semantic predictors in statistical language modeling," in *Proc. Int. Conf. Spoken Language Proc.*, Sydney, Australia, pp. 2403–2406, Dec. 1998.

[49] W. W. Cohen, "Learning rules that classify e-mail," in *Proc. AAAI Spring Symp. Mach. Learning Inf. Access*, Stanford, California, 1996.

[50] A. Conkie and S. Isard, "Optimal coupling of diphones," in *Progress in Speech Synthesis*, J. van Santen, R. Sproat, J. Hirschberg and J. Olive, Eds. New York: Springer, pp. 293–304, 1997.

[51] G. Coorman, J. Fackrell, P. Rutten and B. Van Coile, "Segment selection in the L and H Realspeak Laboratory TTS system," in *Proc. Int. Conf. Spoken Language Proc.*, Beijing, China, pp. 395–398, 2000.

[52] J. K. Cullum and R. A. Willoughby, Lanczos algorithms for large symmetric eigenvalue computations, *Real Rectangular Matrices*, Vol. 1, chapter 5 Boston: Brickhauser, 1985.

[53] W. Daelemans, J. Zavrel, K. van der Sloot and A. van den Bosch, "TiMBL: Tilburg memory based learner, Version 4.0, Reference guide," *ILK Technical Report 01-04*, University of Antwerp, Belgium: Tilburg University and CNTS Research Group, 2001.

[54] A. Dalli, "Biologically inspired lexicon structuring technique," in *Proc. Human Lang. Technol. Workshop*, San Diego, CA, pp. 341–343, March 2002.

[55] R. I. Damper, Y. Marchand, M. J. Adamson and K. Gustafson, "Evaluating the pronunciation component of text-to-speech systems for english: a performance comparison of different approaches," *Computer Speech and Language*, Vol. 13, No. 2, pp. 155–176, 1999.

[56] R. I. Damper, C. Z. Stanbridge and Y. Marchand, "A pronunciation-by-analogy module for the festival text-to-speech synthesiser," in *Proc. 4th Int. Workshop Speech Synthesis*, Pitlochry, Scotland, pp. 97–102, Aug. 2001.

[57] R. De Mori, "Recognizing and using knowledge structures in dialog systems," in *Proc. Aut. Speech Recog. Understanding Workshop*, Keystone, CO, pp. 297–306, Dec. 1999.

[58] S. Deerwester, S. T. Dumais, G. W. Furnas, T. K. Landauer and R. Harshman, "Indexing by latent semantic analysis," *J. Am. Soc. Inf. Sci.*, Vol. 41, pp. 391–407, 1990.

[59] C. H. Q. Ding, "A similarity-based probability model for latent semantic indexing," in *Proc. 22nd Ann. ACM SIGIR Conf. Research and Development Information Retrieval*, Berkeley, CA, pp. 58–65, Aug. 1999.

[60] G. Doddington, "Speaker recognition-identifying people by their voices," *Proc. IEEE*, Vol. 73, Nov. 1985.

[61] R. E. Donovan, "A new distance measure for costing discontinuities in concatenative speech synthesis," in *Proc. 4th ISCA Speech Synthesis Workshop*, Pethshire, Scotland, pp. 59–62, Sept. 2001.

[62] H. Drucker, D. Wu and V. N. Vapnik, "Support vector machines for spam categorization," *IEEE Trans. Neural Networks*, Vol. 10, No. 5, pp. 1048–1054, 1999. 10.1109/72.788645

[63] S. T. Dumais, "Improving the retrieval of information from external sources," *Behav. Res. Methods Instrum. Comput.*, Vol. 23, No. 2, pp. 229–236, 1991.

[64] S. T. Dumais, "Latent semantic indexing (LSI) and TREC-2," in *Proc. Second Text Retrieval Conference (TREC-2)*, D. Harman, Ed., NIST Pub 500–215, Gaithersburg, MD, pp. 105–116, 1994.

[65] T. Dutoit, *An Introduction to Text–to–Speech Synthesis*, Norwell, MA: Kluwer, 1997.

[66] U. M. Fayyad, C. Reina and P. S. Bradley, "Initialization of iterative refinement clustering algorithms," in *Knowledge Discovery and Data Mining*, New York, NY, pp. 194–198, 1998.

[67] M. Federico and R. De Mori, "Language Modeling," in *Spoken Dialogues with Computers*, R. De Mori, Ed., London: Academic, Chapter 7, pp. 199–230, 1998.

[68] P. Flajolet and A. M. Odlyzko, "Random mapping statistics," in *Lecture Notes Comput. Sci.*, Vol. 434, pp. 329–354, 1990.

[69] P. W. Foltz and S. T. Dumais, "Personalized information delivery: an analysis of information filtering methods," *Commun. ACM*, Vol. 35, No. 12, pp. 51–60, 1992. 10.1145/138859.138866

[70] A. Font Llitjos and A. W. Black, "Knowledge of language origin improves pronunciation accuracy of proper names," in *Proc. Eurospeech*, Aalborg, Denmark, pp. 1919–1922, Sept. 2001.

[71] S. Furui, T. Kikuchi, Y. Shinnaka and C. Hori, "Speech-to-text and speech-to-speech summarization of spontaneous speech," *IEEE Trans. Speech Audio Proc.*, Vol. 12, No. 4, pp. 401–408, July 2004. 10.1109/TSA.2004.828699

[72] L. Galescu and J. Allen, "Bi-directional conversion between graphemes and phonemes using a joint N-gram model," in *Proc. 4th ISCA Tutorial and Research Workshop on Speech Synthesis*, Perthshire, Scotland, Aug. 2001.

[73] L. Galescu and J. Allen, "Pronunciation of proper names with a joint N-gram model for bi-directional conversion grapheme-to-phoneme conversion," in *Proc. Int. Conf. Spoken Language Proc.*, Denver, CO, pp. 109–112, Sept. 2002.

[74] P. N. Garner, "On topic identification and dialogue move recognition," *Comput. Speech Lang.*, Vol. 11, No. 4, pp. 275–306, 1997. 10.1006/csla.1997.0032

[75] D. Gildea and T. Hofmann, "Topic-based language modeling using EM," in *Proc. 6th Euro. Conf. Speech Commun. Technol.*, Budapest, Hungary, Vol. 5, pp. 2167–2170, Sept. 1999.

[76] G. Golub and C. Van Loan, *Matrix Computations*, Baltimore, MD: John Hopkins, 2nd edition, 1989.

[77] A. L. Gorin, G. Riccardi and J. H. Wright, "How may i help you?," *Speech Commun.*, Vol. 23, pp. 113–127, 1997. 10.1016/S0167-6393(97)00040-X

[78] Y. Gotoh and S. Renals, "Document space models using latent semantic analysis," in *Proc. 5th Euro. Conf. Speech Commun. Technol.*, Rhodes, Greece, Vol. 3, pp. 1443–1448, Sept. 1997.

[79] A. C. Graesser, K. Wiemer–Hastings, P. Wiemer-Hastings and R. Kreuz, "AutoTutor: a simulation of a human tutor," in *J. Cognit. Syst. Res.*, Vol. 1, pp. 35–51, 1999. 10.1016/S1389-0417(99)00005-4

[80] C. L. Green and P. Edwards, "Using machine learning to enhance software tools for internet information management," in *Proc. AAAI Workshop Internet-Based Inf. Syst.*, Portland, OR, pp. 48–55, 1996.

[81] W. Grei, A. Morgan, R. Fish, M. Richards and A. Kundu, "Fine-grained hidden markov modeling for broadcast-news story segmentation," in *Proc. Human Language Technol. Conf.*, San Diego, CA, 2001.

[82] M. A. Hearst, "TestTiling: segmenting text into multi-paragraph subtopic passages," *Comput. Ling.*, Vol. 23, No. 1, pp. 33–64, 1997.

[83] A. Higgins, L. Bahler and J. Porter, *Digit. Signal Process.*, Vol. 1, pp. 89–106, 1991. 10.1016/1051-2004(91)90098-6

[84] T. Hofmann, "Probabilistic latent semantic analysis," in *Proc. 15th Conf. Uncertainty in AI*, Stockholm, Sweden, July 1999.

[85] T. Hofmann, "Probabilistic topic maps: navigating through large text collections," in *Lecture Notes in Computer Science*, No. 1642, pp. 161–172, Heidelberg: Springer, July 1999.

[86] T. Hofmann, "Probabilistic latent semantic indexing," in *Proc. 22nd Ann. ACM SIGIR Conf. Research and Development Information Retrieval*, Berkeley, CA, pp. 50–57, Aug. 1999.

[87] A. Hunt and A. Black, "Unit selection in a concatenative speech synthesis system using large speech database," in *Proc. Int. Conf. Acoust., Speech, Signal Proc.*, Atlanta, GA, pp. 373–376, 1996.

[88] R. Iyer and M. Ostendorf, "Modeling long distance dependencies in language: topic mixtures versus dynamic cache models," *IEEE Trans. Speech Audio Proc.*, Vol. 7, No. 1, Jan. 1999.

[89] F. Jelinek, "The development of an experimental discrete dictation recognizer," *Proc. IEEE*, Vol. 73, No. 11, pp. 1616–1624, Nov. 1985.

[90] F. Jelinek, "Self-organized language modeling for speech recognition," in *Readings in Speech Recognition*, A. Waibel and K. F. Lee, Eds. San Mateo, CA: Morgan Kaufmann Publishers, pp. 450–506, 1990.

[91] F. Jelinek and C. Chelba, "Putting language into language modeling," in *Proc. 6th Eur. Conf. Speech Commun. Technol.*, Budapest, Hungary, Vol. 1, pp. KN1–KN5, Sept. 1999.

[92] D. Jurafsky, C. Wooters, J. Segal, A. Stolcke, E. Fosler, G. Tajchman and N. Morgan, "Using a stochastic context-free grammar as a language model for speech recognition," in *Proc. 1995 Int. Conf. Acoust., Speech, Signal Proc.*, Detroit, MI, Vol. I, pp. 189–192, May 1995.

[93] P. Kanerva, J. Kristoferson and A. Holst, "Random indexing of text samples for latent semantic analysis," in *Proc. 22nd Ann. Conf. Cognitive Science Society*, L. R. Gleitman

[94] A. K. Kienappel and R. Kneser, "Designing very compact decision trees for grapheme-to-phoneme transcription," in *Proc. Eurospeech*, Aalborg, Denmark, pp. 1911–1914, Sept. 2001.

[95] E. Klabbers and R. Veldhuis, "Reducing audible spectral discontinuities," *IEEE Trans. Speech Audio Proc., Special Issue Speech Synthesis*, N. Campbell, M. Macon and J. Schroeter, Eds. Vol. 9, No. 1, pp. 39–51, Jan. 2001.

[96] D. Klatt, "Review of text to speech conversion for english," *J. Acoust. Soc. Am.*, Vol. 82, pp. 737–793, 1987. 10.1121/1.395275

[97] R. Kneser, "Statistical language modeling using a variable context," in *Proc. Int. Conf. Spoken Language Proc.*, pp. 494–497, Philadelphia, PA, Oct. 1996.

[98] T. Kohonen, S. Kaski, K. Lagus, J. Salojvi, J. Honkela, V. Paatero and A. Saarela, "Self-organization of a massive document collection," *IEEE Trans. Neural Networks*, Vol. 11, No. 3, pp. 674–585, 2000.10.1109/72.846729

[99] E. Kokiopoulou and Y. Saad, "Polynomial filtering in latent semantic indexing for information retrieval," in *Proc. 27th Ann. ACM SIGIR Conf. Research and Development Information Retrieval*, Sheffield, UK, pp. 104–111, July 2004.

[100] T. Kolda and D. P. O'Leary, "Latent semantic indexing via a semi-discrete matrix decomposition," in *The Mathematics of Information Coding, Extraction and Distribution*, G. Cybenko et al., Eds. (*IMA Volumes in Mathematics and Its Applications, Vol. 107*) pp. 73–80, Berlin: Springer, 1999.

[101] L. Lamel, J. L. Gauvain, G. Adda and M. Adda-Decker, "The LIMSI RailTel system: field trial of a telephone service for rail travel information," *Speech Commun.*, Vol. 23, pp. 67–82, 1997. 10.1016/S0167-6393(97)00037-X

[102] T. K. Landauer, D. Laham, B. Rehder and M. E. Schreiner, "How well can passage meaning be derived without using word order: a comparison of latent semantic analysis and humans," in *Proc. Cognit. Sci. Soc.*, pp. 412–417, 1997.

[103] T. K. Landauer and S. T. Dumais, "Solution to Plato's problem: the latent semantic analysis theory of acquisition, induction, and representation of knowledge," *Psychol. Rev.*, Vol. 104, No. 2, pp. 211–240, 1997. 10.1037/0033-295X.104.2.211

[104] P. Langley, I. Wayne, and K. Thompson, "An analysis of Bayesian classifiers," in *Proc. 10th Natl. Conf. on AI*, San Jose, CA, pp. 223–228, 1992.

[105] H. Langseth and T. D. Nielsen, "Classification using hierarchical naive bayes models," Technical Report TR-02-004, Aalborg University, Denmark: Department of Computer Science, 2002.

[106] R. Lau, R. Rosenfeld and S. Roukos, "Trigger-based language models: a maximum entropy approach," in *Proc. 1993 Int. Conf. Acoust., Speech, Signal Proc.*, Minneapolis, MN, pp. II45–48, May 1993.

[107] L.-S. Lee and S. C. Chen, "Automatic title generation for Chinese spoken documents considering the special structure of the language," in *Proc. 8th Euro. Conf. Speech Comm. Technol.*, Geneva, Switzerland, pp. 2325–2328, Sept. 2003.

[108] L.-S. Lee and B. Chen, "Multi-media content understanding and organization based on speech information for efficient indexing/retrieval/browsing applications," *IEEE Signal Process. Mag.*, Vol. 22, No. 5, pp. 42–60, Sept. 2005. 10.1109/MSP.2005.1511823

[109] Q. Li, B.-H. Juang, Q. Zhou and C.-H. Lee, "Automatic verbal information verification for user authentification," *IEEE Trans. Speech Acoust. Proc.*, Vol. 8, No. 5, pp. 585–596, Sept. 2000. 10.1109/89.861378

[110] M.-H. Lin, "Out-of-Core singular value decomposition," *Technical Report TR-83*, Sate University of New York Stony Brook, NY, Experimental Computer Systems Laboratory, May 2000.

[111] R. W. P. Luk and R. I. Damper, "English letter–phoneme conversion by stochastic transducers," in *Data-Driven Techniques in Speech Synthesis*, R. I. Damper, Ed., pp. 91-123, Dordrecht: Kluwer, 2001.

[112] C. X. Ma and M. A. Randolph, "An approach to automatic phonetic baseform generation based on Bayesian networks," in *Proc. Eurospeech*, Aalborg, Denmark, pp. 1453–1456, Sept. 2001.

[113] Y. Marchand and R. I. Damper, "A multi-strategy approach to improving pronunciation by analogy," *Comput. Linguist.*, Vol. 26, No. 2, pp. 195–219, 2001. 10.1162/089120100561674

[114] T. Matsui and S. Furui, "Speaker adaptation of tied-mixture-based phoneme models for text-prompted speaker recognition," in *Proc. 1994 ICASSP*, Adelaide, Australia, pp. 125–128, April 1994.

[115] H. Ney, U. Essen and R. Kneser, "On structuring probabilistic dependences in stochastic language modeling," *Comput. Speech Lang.*, Vol. 8, pp. 1–38, 1994. 10.1006/csla.1994.1001

[116] T. Niesler and P. Woodland, "A variable-length category-based N-gram language model," in *Proc. 1996 Int. Conf. Acoust. Speech, Sig. Proc.*, Atlanta, GA, pp. I164–I167, May 1996.

[117] M. Novak and R. Mammone, "Use of non-negative matrix factorizations for language model adaptation in lecture transcription task," in *Proc. 2001 Int. Conf. Acoust., Speech, Sig. Proc.*, Salt Lake City, UT, Vol 1, pp. 541–544, May 2001.

[118] V. Pagel, K. Lenzo and A. W. Black, "Letter-to-sound rules for accented lexicon compression," in *Proc. Int. Conf. Spoken Language Proc.*, Sydney, Australia, pp. 2015–2018, Dec. 1998.

[119] C. H. Papadimitriou, P. Raghavan, H. Tamaki and S. Vempala, "Latent semantic indexing: a probabilistic analysis," in *Proc. 17th ACM Symp. Princip. Database Syst.*, Seattle, WA, pp. 159–168, June 1998.

[120] S. Parthasaraty and A. E. Rosenberg, "General phrase speaker verification using sub-word background models and likelihood-ratio scoring," in *Proc. Int. Conf. Spoken Language Proc.*, Philadelphia, PA, Oct. 1996.

[121] T. R. Payne and P. Edwards, "Interface agents that learn: an investigation of learning issues in a mail agent interface," *Appl. Artif. Intell.*, Vol. 11, No. 1, pp. 1–32, 1997.

[122] F. C. Pereira, Y. Singer and N. Tishby, "Beyond word *n*-grams," *Comput. Linguist.*, Vol. 22, June 1996.

[123] L. R. Rabiner, B. H. Juang and C.-H. Lee, "An overview of automatic speech recognition," in *Automatic Speech and Speaker Recognition: Advanced Topics*, C.-H. Lee, F. K. Soong and K. K. Paliwal, Eds., Boston, MA: Kluwer, chapter 1 pp. 1–30, 1996.

[124] J. D. M. Rennie, "IFILE: an application of machine learning to e–mail filtering," in *Proc. KDD-2000 Workshop Text Mining*, Boston, MA, 2000.

[125] G. Riccardi and A. L. Gorin, "Stochastic language adaptation over time and state in natural spoken dialog systems," *IEEE Trans. Speech Audio Proc.*, Vol. 8, No. 1, pp. 3–10, Jan. 2000. 10.1109/89.817449

[126] M. Rosen–Zvi, T. Griffiths, M. Steyvers and P. Smyth, "The author-topic model for authors and documents," in *20th Conf. Uncertainty Artificial Intelligence*, Banff, Canada, July 2004.

[127] R. Rosenfeld, "The CMU Statistical language modeling toolkit and its use in the 1994 ARPA CSR evaluation," in *Proc.* ARPA *Speech and Natural Language Workshop*, San Mateo, CA: Morgan Kaufmann Publishers, March 1994.

[128] R. Rosenfeld, "Optimizing lexical and N-gram coverage via judicious use of linguistic data," in *Proc. 4th Euro. Conf. on Speech Commun. Technol.*, Madrid, Spain, pp. 1763–1766, Sept. 1995.

[129] R. Rosenfeld, "A maximum entropy approach to adaptive statistical language modeling," in *Computer Speech and Language*, Vol. 10, London: Academic, pp. 187–228, July 1996. 10.1006/csla.1996.0011

[130] R. Rosenfeld, "Two decades of statistical language modeling: where do we go from here," *Proc. IEEE*, *Special Issue Speech Recog. Understanding*, B. H. Juang and S. Furui, Eds. Vol. 88, No. 8, pp. 1270–1278, Aug. 2000.

[131] M. Sahami, S. Dumais, D. Heckerman and E. Horvitz, "A Bayesian approach to filtering junk e–mail," in *Learning for Text Category (Papers from AAAI Workshop)*, *AAAI Technical Report* WS-98-05, Madison, WI, pp. 55–62, 1998.

[132] G. Sakkis, I. Androutsopoulos, G. Paliouras, V. Karkaletsis, C. D. Spyropoulos and P. Stamatopoulos, "Stacking classifiers for anti-spam filtering of e-mail," in *Proc. 6th Conf. Empir. Methods Nat. Language Proc.*, Pittsburgh, PA, pp. 44–50, 2001.

[133] G. Salton, A. Singhal, M. Mitra and C. Buckley, "Automatic text structuring and summarization," in *Proc. Inf. Process. Manage.*, Vol. 33, No. 2, pp. 193–207, 1997. 10.1016/S0306-4573(96)00062-3

[134] K.-M. Schneider, "A comparison of event models for naive Bayes anti-spam e-mail filtering," in *Proc. 11th Conf. Eur. Chap. ACL*, Budapest, Hungary, 2003.

[135] R. Schwartz, T. Imai, F. Kubala, L. Nguyen and J. Makhoul, "A maximum likelihood model for topic classification of broadcast news," in *Proc. 5th Eur. Conf. Speech Commun. Technol.*, Rhodes, Greece, Vol. 3, pp. 1455–1458, Sept. 1997.

[136] P. Smaragdis, "Discovery auditory objects through non-negativity constraints," in *Proc. ISCA Tutorial Res. Workshop Stat. Perceptual Audio Proc.*, Jeju Island, Korea, Paper 161, Oct. 2004.

[137] J. O. Smith III, "Why sinusoids are important," in *Mathematics of the Discrete Fourier Transform (DFT)*, W3K Publishing, 2003, Section 4.1.2 ISBN 0-9745607-0-7.

[138] Speech Assessment Methods Phonetic Alphabet (SAMPA), "Standard machine-readable encoding of phonetic notation," *ESPRIT Project 1541, 1987–89*, cf. http://www.phon.ucl.ac.uk/home/sampa/home.htm.

[139] E. Spertus, "Smokey: automatic recognition of hostile messages," in *Proc. 14th Natl. Conf. on AI and 9th Conf. Innov. Appl. of AI*, Providence, RI, pp. 1058–1065, 1997.

[140] R. E. Story, "An explanation of the effectiveness of latent semantic indexing by means of a Bayesian regression model," *Inf. Process. & Manag.*, Vol. 32, No. 3, pp. 329–344, 1996. 10.1016/0306-4573(95)00055-0

[141] Y. Stylianou, "Removing phase mismatches in concatenative speech synthesis," in *Proc. 3rd ESCA Speech Synthesis Workshop*, Jenolan Caves, Australia, pp. 267–272, Nov. 1998.

[142] Y. Stylianou and A. K. Syrdal, "Perceptual and objective detection of discontinuities in concatenative speech synthesis, in *Proc. Int. Conf. Acoust., Speech, Signal Proc.*, Salt Lake City, UT, pp. 837–840, 2001.

[143] J. Suontausta and J. Häkkinenen, "Decision tree based text-to-phoneme mapping for speech recognition," in *Proc. Int. Conf. Spoken Language Proc.*, Beijing, China, pp. 831–834, Oct. 2000.

[144] K. Takeda, K. Abe and Y. Sagisaka, "On the basic scheme and algorithms in nonuniform unit speech synthesis," in *Talking Machines*, G. Bailly and C. Benoit, Eds. Amsterdams: North–Holland, pp. 93–105, 1992.

[145] Y. Teh, M. Jordan, M. Beal and D. Blei, "Hierarchical dirichlet processes," in *Advances Neural Information Processing Systems*, Vancouver, Canada, 2004.

[146]   M. Tsuzaki and H. Kawai, "Feature Extraction for unit selection in concatenative speech synthesis: comparison between AIM, LPC, and MFCC," in *Proc. Int. Conf. Spoken Language Proc.*, Denver, CO, pp. 137–140, Sept. 2002.

[147]   J. Vepa and S. King, "Join cost for unit selection speech synthesis," in *Text to Speech Synthesis: New Paradigms and Advances*, S. Narayanan and A. Alwan, Eds. Upper Saddle River, NJ: Prentice Hall, pp. 35–62, 2004.

[148]   M. Vingron, "Near-optimal sequence alignment," *Curr. Opin. Struct. Biol.*, Vol. 6, No. 3, pp. 346–352, June 1996. 10.1016/S0959-440X(96)80054-6

[149]   Y.-Y. Wang, L. Deng and A. Acero, "An introduction to statistical spoken language understanding," *IEEE Signal Process. Mag.*, Vol. 22, No. 5, pp. 16–31, Sept. 2005. 10.1109/MSP.2005.1511821

[150]   M. J. Witbrock and V. O. Mittal, "Ultra-summarization: a statistical approach to generating highly condensed non-extractive summaries," in *Proc. ACM SIGIR Int. Conf. R and D Inf. Retr.*, Berkeley, CA, pp. 315–316, Aug. 1999.

[151]   J. Wouters and M. W. Macon, "A perceptual evaluation of distance measures for concatenation speech synthesis," in *Proc. Int. Conf. Spoken Language Proc.*, Sydney, Australia, Vol. 6, pp. 159–163, Dec. 1998.

[152]   F. Yvon, "Pradigmatic cascades: a linguistically sound model of pronunciation by analogy," in *Proc. 35th Ann. Meeting Assoc. Comput. Linguit.*, pp. 428–435, 1997.

[153]   V. Zue, S. Seneff, J. R. Glass, J. Polifroni, C. Pao, T. J. Hazen and L. Hetherington, "JUPITER: a telephone-based conversational interface for weather information," *IEEE Trans. Speech Audio Proc.*, Vol. 8, No. 1, pp. 85–96, Jan. 2000. 10.1109/89.817460

Author Biography

Jerome R. Bellegarda received the Diplôme d'Ingénieur degree (summa cum laude) from the Ecole Nationale Supérieure d'Electricité et de Mécanique, Nancy, France, in 1984, and the M.S. and Ph.D. degrees in Electrical Engineering from the University of Rochester, Rochester, NY, in 1984 and 1987, respectively.

From 1988 to 1994, he was a Research Staff Member at the IBM T.J. Watson Research Center, Yorktown Heights, NY, working on speech and handwriting recognition, particularly acoustic and chirographic modeling. In 1994, he joined Apple Inc., Cupertino, CA, where he is currently Apple Distinguished Scientist in Speech & Language Technologies. At Apple he has worked on many facets of human language processing, including speech recognition, speech synthesis, statistical language modeling, voice authentication, speaker adaptation, dialog interaction, metadata extraction, and semantic classification. In these areas he has written close to 150 journal and conference papers, and holds over 30 patents. He has also contributed chapters to several edited books, most recently *Pattern Recognition in Speech and Language Processing* (New York, NY: CRC Press, 2003), and *Mathematical Foundations of Speech and Language Processing* (New York, NY: Springer-Verlag, 2004). His research interests include statistical modeling algorithms, voice-driven man-machine communications, multiple input/output modalities, and multimedia knowledge management.

Dr. Bellegarda has served on many international scientific committees, review panels, and editorial boards. Between 2000 and 2004, he was Associate Editor of the *IEEE Transactions on Speech and Audio Processing*, and served as a member of the Speech Technical Committee (STC) of the IEEE Signal Processing Society (SPS). Since 2004, he has been on the Editorial Board of *Speech Communication*. In 2003, he was elected Fellow of the IEEE "for contributions to statistical methods for human language processing." He was the STC nominee for both the 2001 IEEE W.R.G. Baker Prize Paper Award and the 2003 IEEE SPS Best Paper Award with the paper "Large Vocabulary Speech Recognition With Multi-Span Statistical Language Models" published in *IEEE Transactions on Speech and Audio Processing* (Piscataway, NJ: IEEE Press, 2000). He was recently awarded the 2006 International Speech Communication Association Best Paper Award for the paper "Statistical Language Model Adaptation: Review and Perspectives," published in *Speech Communication* (Amsterdam, The Netherlands: Elsevier Science, 2004).